Plain-English Study Guide for the FCC Amateur Radio Technician Class License

Richard P. Clem, WØIS

Illustrated by

Yippy G. Clem, KCØOIA

2022-2026 Edition

Copyright © 2026 Richard P. Clem and Yippy G. Clem

All rights reserved.

Richard P. Clem
P.O. Box 14957
Minneapolis, MN 55414
ISBN:

CONTENTS

	Introduction	v
1	Basic Electricity and Electronics	7
2	Radio Waves	53
3	Antennas	62
4	Propagation of Radio Waves	74
5	Radio Equipment and Setting Up a Station	82
6	Different Modes	92
7	Different Operating Activities	107
8	Operating Procedures	115
9	FCC Rules	124
10	Safety	138
11	Miscellaneous	148

INTRODUCTION

This study guide will prepare you to take the FCC Technician Class Amateur radio exam. After you read this guide, you will be able to pass the test with no difficulty.

Many persons have been able to pass the test merely by memorizing the answers to the questions. While that is one way to pass the test, it is not recommended. First of all, the test is not particularly difficult. And while it might be possible to memorize the almost 400 questions, that is actually a much more daunting task than simply learning the material.

The questions in this guide are the same questions that will be asked on the test. You will be given 35 of these questions. To pass the test, you will need to answer 26 of those questions correctly. There is no penalty for wrong answers, so if you don't know the answer, the best strategy is to guess. You can almost always eliminate one or more wrong answers, which will increase your odds of guessing the right answer. The questions included here will be used through June 30, 2026. Starting July 1, 2026, there will be a new version of the test, and a new version of this guide will be published.

This guide will provide an explanation of the material, in plain English. In a few cases, I will probably be guilty of oversimplification. But again, the Technician exam is not particularly difficult, and you can easily pass the test with just a superficial understanding of the material. After you get on the air, many of the concepts explained here will become much more apparent to you. Even if you initial understanding is superficial, you will soon learn more just by doing.

The exam itself was never intended to prove that people who pass it are experts in the field of electronics or radio. Particularly at the Technician level, it was designed to show that you have a basic understanding of the material, so that you will be able to learn more once you get on the air. This little book will both give you that basic understanding, and also give you the knowledge to easily pass the test.

Each section of this book contains an explanation of some of the material that is covered on the test. This is followed by the actual test questions. The correct answers are **underlined**. In most cases, after you read the section, you will be able to easily answer the questions. In a few cases, you might need to go back and re-read that section.

There might still be a couple of occasions when you still can't answer the questions, even after re-reading the section. I guarantee you that those occasions will be few. But if they do arise, you might have a handful of questions where you should simply memorize the question and the answer. To pass the test, you merely have to supply the correct answer. If you can do that by memorization, you will still get the question right, and you will still pass the test. **If you do memorize, you must memorize the correct answer, not just the letter for the answer.** The questions on the test will be exactly as shown here. However, the possible answers will probably be in a different order.

Don't attempt to memorize all of the answers. For one thing, that will be overwhelming. I know that I couldn't possibly memorize 400 questions and answers. And you'll get more out of the entire process, and have more fun on the air, if you have a basic understanding of the material. But some concepts will be new to you, and you might not have a good understanding. For those handful of times when you just don't "get it", don't feel bad if you have to simply memorize the correct answer.

1 BASIC ELECTRICITY AND ELECTRONICS

As you probably know, everything in the universe is made of a basic building block called the atom. The atom, in turn, is made up of three infinitesimally small particles called protons, neutrons, and electrons. The protons and neutrons are located in the center of the atom, and typically never go anywhere. The electrons, on the other hand, are orbiting the center (like tiny planets around the sun) and can move about much more freely.

Each chemical element has a certain number of protons, a certain number of neutrons, and a certain number of electrons. Since the electrons are free to move about, they sometimes leave the atom and go elsewhere. Or, an extra electron might enter the atom. Therefore, there are some atoms that have too many electrons, or two few electrons.

Typically, an atom that has too many electrons is eager to get rid of the extra atom, and it will do its best to get rid of the intruder. Given the opportunity, it will try to unload its extra electron on another unsuspecting atom nearby, as if it were a hot potato.

Similarly, if an atom has too few electrons, it wants to find a replacement wherever it can, and will snatch an electron away from another nearby atom, given the opportunity.

Picture in your mind a chain of atoms, sitting side by side. Let's say that the atom on the far left has an extra electron that it doesn't want. And let's say that the atom on the right is missing one of its electrons.

Realizing that it has an extra electron, the atom on the left waits until the atom to the right isn't looking, and then hands the "hot potato" electron to its neighbor. This atom doesn't want a spare electron any more than the original atom did, so it does its best to get rid of it. The original atom, having started this whole process in the first place, is obviously aware of what is going on, so the second atom probably won't be too successful in giving it back.

So the second atom waits until an opportune moment, and then gives the electron to the atom sitting to its right.

This whole process is repeated until the extra electron has been passed all the way down the line. But the last atom, the one at the right, was missing an electron. So this atom is quite happy to get the replacement, and all is well again in the atomic world.

This whole process is what is called an electric current. There was an electric current flowing from the atom on the left, all the way to the atom on the right.

There is a measurement of how strong an electric current is, and this measurement is the **Ampere**, or **Amp** for short. One Amp is defined as approximately 6000000000000000000 (that is 6, followed by 18 zeros) electrons per second passing a certain point. (You do not need to know that number for the test—it's just to give you an idea that it's a **lot** of electrons.)

It is sometimes useful to think of electricity as being similar to water. Electricity consists of electrons flowing through a wire. For many purposes, this is the similar to water flowing through a pipe. Therefore, you can think of the electric current, in amps, as being similar to the number of gallons per second passing through a pipe: It's a measure of how fast the electricity is moving.

Electrical current is measured in which of the following units?
A. Volts
B. Watts
C. Ohms
D. Amperes

What is the name for the flow of electrons in an electric circuit?
A. Voltage
B. Resistance
C. Capacitance
D. Current

Let's think about that water pipe, because the behavior of water flowing through a pipe is similar to the behavior of electricity flowing through a wire. If we want to know what size of pipe to order, we'll want to know how much water is flowing through it—how many gallons per minute. If there is more flow, we'll need a bigger pipe.

Similarly, if there is more electric current (more amps), we'll need a bigger wire to handle the current.

But when we put in the new pipe, we'll also need to know how thick the walls of the pipe need to be. If there is too much water pressure and the pipe is made out of flimsy material, then the pipe might burst. We need to know how much pressure there will be. Normally, this is measured in pounds per square inch. This figure tells us how much energy there is in each drop of water.

There is a similar concept in electricity, and that is **voltage**. Voltage, as you might have guessed, is measured in **volts**. The number of volts tells us how much energy there is in each of the electrons that is passing through the wire. Just because we have a high current does not mean that the voltage will be high. This is just like water. If you look at a river, there is a huge amount of water passing by, so the current is very high. But the pressure is very low. (But if you try to squeeze that same river into a small pipe, the pressure will get much higher, a concept we'll talk about in a few minutes.)

Another name for voltage is **electromotive force**. The higher the voltage, the more energy there is in each electron–the more pressure there is.

What is the electrical term for the force that causes electron flow?

A. Voltage
B. Ampere-hours
C. Capacitance
D. Inductance

Let's think about that water pipe again. The water pipe running down the street in front of your house is probably a foot or more in diameter, because it is supplying water to all of the houses on your street. A lot of water passes through it, so it needs to be larger. The pipe running into your house has a smaller diameter, because less water needs to pass through it. The pipes running to the individual sinks in your house are even smaller, because they need to carry even less water.

If you tried to replace the pipe in front of your house with the small pipe, it would cause much resistance to the flow of water. There is a similar concept in electricity, called **resistance**. Resistance is the measurement of how much a particular substance blocks the flow of electricity. Substances with a very low resistance (like the big pipe running in the street) are good at conducting electricity. We call these substances **conductors**. In general, metals are good conductors of electricity–they have very low resistance. The reason for this is because they have many free electrons. Non-metallic substances, such as plastic, dry wood, rubber, etc., have an extremely high resistance (for most practical purposes, they have infinite resistance). These substances are called **insulators**.

Gold and silver are among the best electrical conductors. Copper and aluminum follow close behind, and they are the most common substances used for electric wires. Other metals are generally fairly good conductors. A piece of these materials would generally have a resistance of close to zero.

Some other substances can conduct electricity to some extent, but not as well as a true conductor. For example, carbon will conduct electricity, but it is like a very small pipe, because it doesn't conduct electricity very well. It has a high resistance, but not as high as plastic or rubber. Water can also conduct electricity, particularly if it is salt water. But it is not nearly as good a conductor as copper or other metals.

The amount of resistance of a piece of material is measured in **Ohms**. A perfect conductor will have a resistance of zero ohms. For example, a piece of copper wire will have a resistance very close to zero ohms. An insulator, such as plastic, will have a very high resistance. This will be many millions of ohms, and for practical purposes, we can just say that the resistance is an infinite number of ohms.

When we assemble an electronic device, sometimes we need to introduce resistance to the circuit. We use an electronic component called a **resistor**. When you buy a resistor, you specify the number of ohms that you need. Some resistors will be one ohm or less, and others will be many million ohms. They are generally made out of carbon.

One special kind of resistor is called a potentiometer. This is sometimes called a "variable resistor", but the word "potentiometer" is the one used on the test. A potentiometer is a resistor whose resistance can be changed, usually by turning the knob. If you turn the knob all the way one direction, the resistance is low. If you turn the knob all the way in the other direction, the resistance is high. A potentiometer is usually the component used as a volume control on a radio receiver. With the volume turned all the way down, the resistance is high. With the volume turned all the way up, the resistance is low. But when you turn the knob, you are adjusting the resistance (the number of ohms) of the resistor/potentiometer.

For one question, you need to know that resistance blocks the flow of any kind of current—DC, AC, and RF. (We'll explain later what these stand for.)

Why are metals generally good conductors of electricity?
A. They have relatively high density
B. They have many free electrons
C. They have many free protons
D. All these choices are correct

What are the units of electrical resistance?
A. Siemens
B. Mhos
C. Ohms
D. Coulombs

Which of the following is a good electrical insulator?
A. Copper
B. Glass
C. Aluminum
D. Mercury

What electrical component opposes the flow of current in a DC circuit?
A. Inductor
B. Resistor
C. Inverter
D. Transformer

What type of component is often used as an adjustable volume control?
A. Fixed resistor
B. Power resistor
C. Potentiometer

D. Transformer

What electrical parameter is controlled by a potentiometer?
A. Inductance
B. Resistance
C. Capacitance
D. Field strength

What type of current flow is opposed by resistance?
A. Direct current
B. Alternating current
C. RF current
D. All these choices are correct

Think once again of a pipe with water flowing through it. Let's assume that we have water flowing through a three-inch pipe. At some point, the pipe narrows, and flows into a one-inch pipe.

The current flowing through the pipe is the same everywhere. If you have ten gallons per minute flowing through the big pipe, and the only place for it to go is into the smaller pipe, then you'll have ten gallons per minute flowing through the smaller pipe.

But the water in the one-inch pipe will have more pressure. You can see this from the following diagram:

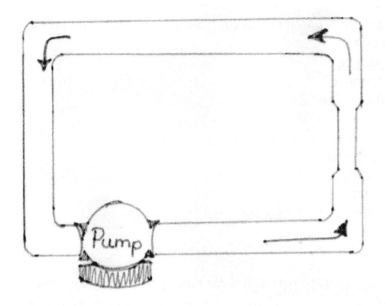

The pump is pumping water through the system. There is ten gallons per minute flowing through the entire loop. But the one-inch pipe is under a lot more pressure than the three-inch pipe.

Electricity works the same way. Look at the diagram below. This is a battery, some wire, and a light bulb. As you are probably aware, the light bulb will light up when it's hooked up this way. The light bulb gives off light because it is using energy from the electricity. And because it is using energy, it resists the flow of electricity. Therefore, it is not a true conductor, and it is not a true insulator. Instead, it is a resistor. We can measure the resistance of the light bulb with an instrument called an **ohmmeter**. For example, this light bulb might have a resistance of 2 ohms:

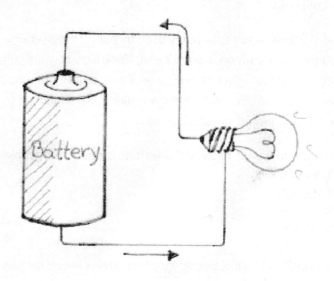

The light bulb in this diagram is like the one-inch pipe in the first diagram. Like the smaller section of pipe, it resists the flow of electricity. The battery in this diagram is like the pump in the first diagram. In the first diagram, the pump might be rated in terms of how much pressure it can generate—how many pounds per square inch. The battery is the same—it is rated in terms of how much electrical pressure (voltage) it puts out. Normal flashlight batteries (alkaline or carbon zinc) put out 1.5 volts. (We'll talk later about how batteries with larger voltages are put together).

As you are probably aware, normal alkaline or carbon-zinc batteries are not rechargeable. Some batteries are rechargeable. The most common ones for small electronic devices are **nickel-cadmium** (also known as **NiCd** or "**nicad**" batteries) or **lithium-ion** (also known as **LiIon** batteries). The voltage of a nicad battery is 1.2 volts.

A normal car battery is known as a "lead-acid" battery. As you are aware, this type of battery can also be recharged, and is normally charged by your car's alternator. Some lead acid batteries are called "gel cels." This just means that the acid inside the battery is in gel form instead of liquid.

Lead-acid batteries are commonly used by hams to power equipment, but you must use caution around them, since they contain sulfuric acid. In addition, they can give off explosive hydrogen gas when being charged. They can overheat, or even explode, particularly if they are charged or discharged too quickly.

Which of the following battery chemistries is not rechargeable?
A. Nickel-cadmium
B. Carbon-zinc
C. Lead-acid
D. Lithium-ion

Which of the following battery y chemistries is rechargeable?
A. Nickel-metal hydride
B. Lithium-ion
C. Lead-acid gel-cell
D. All of these choices are correct

Which of the following is a safety hazard of a 12-volt storage battery?
A. Touching both terminals with the hands can cause electrical shock
B. Shorting the terminals can cause burns, fire, or an explosion
C. RF emissions from a nearby transmitter can cause the electrolyte to emit poison gas
D. All of these choices are correct

What hazard is caused by charging or discharging a battery too quickly?
A. Overheating or out-gassing
B. Excess output ripple
C. Half-wave rectification
D. Inverse memory effect

There is a relationship between voltage, current, and resistance (volts, amps, and ohms). This is a mathematical formula called **Ohm's Law**:

Volts equals amps time ohms.

Even if you don't understand this formula, you will be able to pass the test simply by memorizing the formula. If you are good with algebra, if you remember that first formula, you will be able to figure out the other two versions of the formula. But if you're not good with algebra, you can simply memorize the following two other versions of the formula:

Amps equals volts divided by ohms.

Ohms equals volts divided by amps.

Some of the questions might ask you for the correct formula. On these occasions, the question will use *E* for voltage (short for electromotive force), *I* for current (short for intensity) and *R* for resistance. If you get one of these questions, you need to pick out the correct formula, and these are the correct answers:

$$I = E / R$$

$$E = I \times R$$

$$R = E / I$$

When you see an Ohm's law question on the test, two numbers will be given, and you'll need to calculate the third. So the question might give numbers for amps and ohms. In that case, you'll just multiply the two numbers and get the correct answer in volts. Or the answer might give numbers for volts and ohms, and you'll need to get the number of amps by calculating volts divided by ohms. Or, the question will give volts and amps, and you'll need to give the answer of the number of ohms, which is volts divided by amps.

You can answer all of the questions in this way without understanding anything, as long as you memorize these three formulas. But it doesn't take much extra effort to understand what's going on!

Think again about the water flowing through the first diagram above.

There's a certain amount of current (gallons per minute) flowing through the whole system. But when the water encounters the resistance of the small section of pipe, the pressure in that section goes up. We could assign a number for the resistance. Let's call it 10 "ohms" (of course, resistance of a water pipe isn't really measured in ohms, but it's close enough for the purposes of this exercise. And then, we'll replace that section of pipe with a little bit larger one that is only five "ohms". So we have the following two setups:

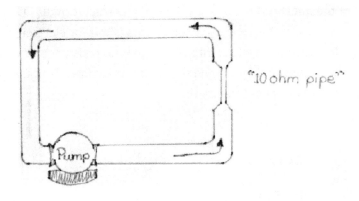

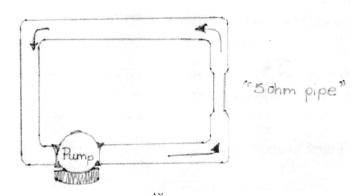

Under "Ohm's Law" (which doesn't really apply here, but again, it's close enough to demonstrate the point), the "voltage" (water pressure) equals current times ohms.

So in the drawing with the "10 ohm pipe", let's say that the current is 10 gallons per minute. The pressure (voltage) would be current times resistance, or 10 x 10 = 100. (At this point, I'm not saying 100 of what, because that would obviously depend on a lot of factors. But for the sake of simplicity, we'll say that the pressure is 100 pounds per square inch.

In the other diagram, we replace the "10 ohm" pipe with the larger "5 ohm" pipe. This time, the pressure (voltage) will be current times resistance, or 10 x 5 = 50.

In other words, when the resistance goes down, the voltage also goes down.

Now, obviously, in the case of water, this is a gross over-simplification, and things don't work out quite this simply. But you get the idea: As long as everything else stays the same, then the voltage goes up when the resistance goes up. This might seem backwards to you when it comes to electricity. But it makes more sense when you think about the water pipes. And it also makes more sense if you think of it this way: The larger resistor "uses up" more of the voltage than the smaller resistor. Or you need more voltage in the case of a larger resistor. We usually say that there is a "voltage drop" across the resistor.

Before we go on to draw the same diagram using electrical components, we should introduce the symbols for various electronic components. My picture of the light bulb and battery above is adequate, but if we want to draw more complicated electrical circuits, my artistic abilities are limited, and it would be very difficult for me to draw everything I need to draw. So we use standard symbols to represent everything. These symbols are called **schematic** symbols, and the layout of a circuit using these symbols is called a **schematic** diagram.

Here is the same diagram we drew earlier, using pictures of a battery and light bulb:

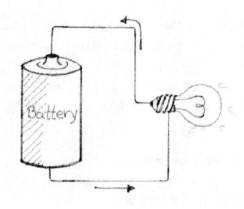

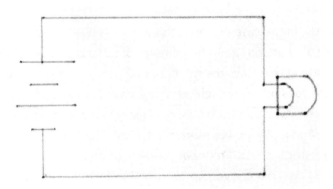

As you can see, we now know the symbols for a light bulb and for a battery:

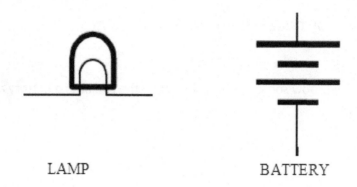

LAMP   BATTERY

And here is the symbol for a resistor. As you can see, it is just a zig-zag line:

RESISTOR

We previously talked about a potentiometer, also known as a variable resistor. Here is the schematic symbol for a variable resistor:

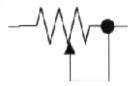

POTENTIOMETER (VARIABLE RESISTOR)

Now that we know some symbols, let's draw the circuit of a battery and a resistor:

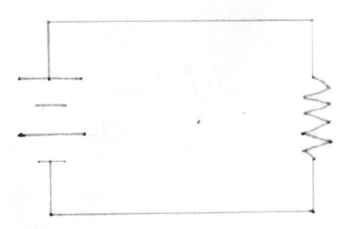

There are three things we can measure in this diagram. We can measure the voltage of the battery (the number of volts). We can measure the current flowing through the wire (the number of amps). And we can measure the resistance of the resistor (the number of ohms).

Using Ohm's law, as long as we know two of these numbers, we can always figure out the missing one.

For example, let's assume that the battery is 12 volts (such as a lead-acid car battery), and the resistor is 4 ohms. We can figure out the current in amps. Remember the formula:

Amps equals volts divided by ohms.

So here, there would be 3 amps, because 12 divided by 4 is 3.

For our next example, let's assume that the battery is 9 volts and the current is 4.5 amps. What is the resistance? Here, we remember the formula:

Ohms equals volts divided by amps.

So the correct answer is 2 ohms: 9 divided by 4.5.

Finally, we know that the current is 2 amps, and the resistance is 6 ohms. What is the voltage? Here, we remember the formula:

Volts equals amps times ohms.

So the battery is 12 volts: 2 times 6 equals 12.

What is the name of an electrical wiring diagram that uses standard component symbols?
A. Bill of materials
B. Connector pinout
C. Schematic
D. Flow chart

The following questions all refer to this diagram:

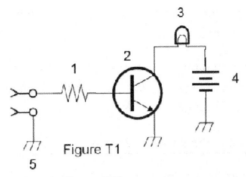

Figure T1

What is component 1 in figure T1?

A. Resistor
B. Transistor
C. Battery
D. Connector

What is component 3 in figure T1?
A. Resistor
B. Transistor
C. Lamp
D. Ground symbol

What is component 4 in figure T-1?
A. Resistor
B. Transistor
C. Ground symbol
D. Battery

The following question refers to this diagram:

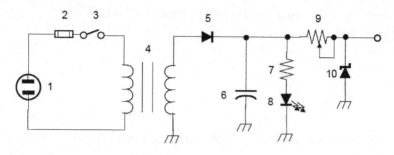

Figure T2

What is component 9 in figure T2?

A. Variable capacitor
B. Variable inductor
C. Variable resistor
D. Variable transformer

What do the symbols on an electrical circuit schematic diagram represent?
A. Electrical components
B. Logic states
C. Digital codes
D. Traffic nodes

Which of the following is accurately represented in electrical schematics?
A. Wire lengths
B. Physical appearance of components
C. Component connections
D. All these choices are correct

What formula is used to calculate current in a circuit?
A. I = E × R
B. I = E / R
C. I = E + R

D. I = E - R

What formula is used to calculate voltage in a circuit?
A. E = I x R
B. E = I / R
C. E = I + R
D. E = I - R

What formula is used to calculate resistance in a circuit?
A. R = E x I
B. R = E / I
C. R = E + I
D. R = E - I

What is the resistance of a circuit in which a current of 3 amperes flows through a resistor connected to 90 volts?
A. 3 ohms
B. 30 ohms
C. 93 ohms
D. 270 ohms

What is the resistance in a circuit for which the applied voltage is 12 volts and the current flow is 1.5 amperes?
A. 18 ohms
B. 0.125 ohms
C. 8 ohms
D. 13.5 ohms

What is the resistance of a circuit that draws 4 amperes from a 12-volt source?
A. 3 ohms
B. 16 ohms
C. 48 ohms
D. 8 Ohms

What is the current in a circuit with an applied voltage of 120 volts and a resistance of 80 ohms?
A. 9600 amperes
B. 200 amperes
C. 0.667 amperes
D. 1.5 amperes

What is the current through a 100-ohm resistor connected across 200 volts?
A. 20,000 amperes
B. 0.5 amperes
C. 2 amperes
D. 100 amperes

What is the current through a 24-ohm resistor connected across 240 volts?
A. 24,000 amperes
B. 0.1 amperes
C. 10 amperes
D. 216 amperes

What is the voltage across a 2-ohm resistor if a current of 0.5 amperes flows through it?
A. 1 volt
B. 0.25 volts
C. 2.5 volts
D. 1.5 volts

What is the voltage across a 10-ohm resistor if a current of 1 ampere flows through it?
A. 1 volt
B. 10 volts
C. 11 volts
D. 9 volts

What is the voltage across a 10-ohm resistor if a current of 2 amperes flows through it?
A. 8 volts
B. 0.2 volts
C. 12 volts
D. 20 volts

A battery supplies electricity that is **direct current**. Direct current (DC) electricity is electricity that flows in one direction all the time. The electrons flow from the negative terminal of the battery, through the circuit, and back to the positive terminal of the battery. They never change directions. (Some people think of electricity as moving from the positive terminal to the negative terminal. It's moving so fast that it really doesn't matter which direction it's moving. The important thing is that DC is always moving the same direction.)

The other type of electricity is **alternating current**. Alternating current (AC) electricity changes direction on a regular basis. The outlets in your home are alternating current electricity. The electricity in your house changes direction 120 times per second. This is called 60 cycles per second, or 60 **Hertz** (abbreviated 60 Hz) electricity, because it goes through 60 complete cycles (it goes both directions) every second. (The unit of frequency is the Hertz.)

For alternating current, there is a concept called **impedance**, which is similar to resistance. It is the amount of opposition to the flow of AC. It is also measured in ohms.

Which of the following describes alternating current?
A. Current that alternates between a positive direction and zero
B. Current that alternates between a negative direction and zero
C. Current that alternates between positive and negative directions
D. All these answers are correct

What describes the number of times per second that an alternating current makes a complete cycle?
A. Pulse rate
B. Speed
C. Wavelength
D. Frequency

What is the unit of frequency?
A. Hertz
B. Henry
C. Farad
D. Tesla

What is impedance?
A. The opposition to AC current flow
B. The inverse of resistance
C. The Q or Quality Factor of a component
D. The power handling capability of a component

What is the unit of impedance?
A. The volt
B. The ampere
C. The coulomb
D. The ohm

Two words that you need to know for the test are **series** and **parallel**. **Series** means that two things are hooked up "end to end" in a straight line. Since all of the current goes through all of the components, you need to remember that the current is the same in all of them. The voltage, on the other hand, is divided up between the components. The exact amount depends on the value of the components.

If two components are hooked up in **parallel**, this means that they are connected side by side. For example, if you connect two batteries in parallel, you hook both of the positive connections together, and both of the negative connections together. You need to remember that the voltage is always the same across components hooked up in parallel. The current, on the other hand, splits in two. Some of the current goes through one component, and the rest of the current goes through the other one. In other words, the current gets divided. The exact amount depends on the value of the components.

In which type of circuit is DC current the same through all components?
A. Series
B. Parallel
C. Resonant
D. Branch

In which type of circuit is voltage the same across all components?
A. Series
B. Parallel
C. Resonant
D. Branch

For the exam, you will need to be familiar with some of the metric prefixes. You will need to know **kilo-, Mega-, Giga-, milli-,** and **micro-,** and **pico-**. You are probably already familiar with these, with the possible exception of the last one. A kilometer, for example, is a thousand meters. A kilogram is a thousand grams. "Megabucks" means a lot of money, in particular, one million dollars. A millimeter is 1/1000 of a meter.

"Pico" is a millionth of a millionth, or 1,000,000,000,000.

Here is the meaning of the ones that you will encounter on the test:

    kilo =   1000 of anything

    Mega =   1,000,000 of anything

Giga = 1,000,000,000 of anything

milli = 1/1000 of anything

micro = 1/1,000,000 of anything

pico = 1/1,000,000,000,000 of anything

For example, a kilowatt is 1000 watts. 4 Megohms is 4,000,000 ohms. One millivolt is 1/1000 volt. One microfarad is 1/1,000,000 Farad. (Don't worry about the exact meaning of these terms at this point. For now, we're just concerned about understanding the meaning of these prefixes.)

There is only one question with the prefix "Giga", which is sometimes used for very high frequencies. You just need to know that 1 GHz = 1000 MHz. For the one question, you need to know that 2425 MHz is the same as 2.425 GHz.

For two questions, it's important to remember that the abbreviation for "megahertz" has a capital M and a capital H. It's a capital M because a small m would mean "milli" rather than mega. And Hertz is a name, so it is capitalized. And for kilohertz, the correct abbreviation is kHz, with a small k and a capital H. Remember that the correct abbreviations are **MHz and kHz.** If you forget this, look through some of the other questions and answers on the test. They will be written the right way, even in the wrong answers.

How many milliamperes is 1.5 amperes?
A. 15 milliamperes
B. 150 milliamperes
C. 1,500 milliamperes
D. 15,000 milliamperes

Which is equal to 1,500,000 hertz?
A. 1500 kHz
B. 1500 MHz
C. 15 GHz

D. 150 kHz

What is the abbreviation for megahertz?
A. MH
B. mh
C. Mhz
D. MHz

What is the abbreviation for kilohertz?
A. KHZ
B. khz
C. khZ
D. kHz

Which is equal to one kilovolt?
A. One one-thousandth of a volt
B. One hundred volts
C. One thousand volts
D. One million volts

Which is equal to one microvolt?
A. One one-millionth of a volt
B. One million volts
C. One thousand kilovolts
D. One one-thousandth of a volt

Which is equal to 500 milliwatts?
A. 0.02 watts
B. 0.5 watts
C. 5 watts
D. 50 watts

Which is equal to 3000 milliamperes?
A. 0.003 amperes
B. 0.3 amperes
C. 3,000,000 amperes
D. 3 amperes

Which is equal to 3.525 MHz?

A. 0.003525 kHz
B. 35.25 kHz
C. 3525 kHz
D. 3,525,000 kHz

Which is equal to 1,000,000 picofarads?
A. 0.001 microfarads
B. 1 microfarad
C. 1000 microfarads
D. 1,000,000,000 microfarads

Which is equal to 28400 kHz?
A. 28.400 kHz
B. 2.800 MHz
C. 284.00 MHz
D. 28.400 MHz

Which is equal to 2425 MHz?
A. 0.002425 GHz
B. 24.25 GHz
C. 2.425 GHz
D. 2425 GHz

Power is the rate at which electrical energy (or any other kind of energy) is being used. In general, when electricity passes through a resistance, it changes into heat. So power is also the rate at which the resistor is giving off heat (or perhaps light, or radio waves, or whatever the device is designed to do). Power is measured in **watts**. So a 100 watt light bulb uses electric energy (and gives of light and heat) twice as fast as a 50 watt light bulb.

Power is equal to voltage times current. Or, to put it another way:

> **Watts equals volts times amps.**

If you know a little bit of algebra, you can figure out the other two versions of this equation:

Volts equals watts divided by amps.

Amps equals watts divided by volts.

As with Ohm's law, any time you know two of these numbers (watts, amps, or volts), you can figure out the other one.

There is one question on the test that asks you to identify the formula, and it is written this way:

$$P = I \times E$$

Electrical power is measured in which of the following units?
A. Volts
B. Watts
C. Ohms
D. Amperes

Which term describes the rate at which electrical energy is used?
A. Resistance
B. Current
C. Power
D. Voltage

What is the formula used to calculate electrical power (P) in a DC circuit?
A. P = I × E
B. P = E / I
C. P = E − I
D. P = I + E

How much power is delivered by a voltage of 13.8 volts DC and a current of 10 amperes?
A. 138 watts
B. 0.7 watts
C. 23.8 watts
D. 3.8 watts

How much power is delivered by a voltage of 12 volts DC and a current of 2.5 amperes?
A. 4.8 watts
B. 30 watts
C. 14.5 watts
D. 0.208 watts

How much current is required to deliver 120 watts at a voltage of 12 volts DC?
A. 0.1 amperes
B. 10 amperes
C. 12 amperes
D. 132 amperes

A **deciBel** (dB) is a unit used to measure the difference in the strength of two things. If you remember the following facts, you will be able to answer all of the questions on the test:

If a signal doubles, that is a difference of approximately 3 dB. If the signal doubles twice, that is 6 dB. For example, if a signal goes from 5 watts to 10 watts, that is 3 dB. If it then goes to 20 watts, that is 3 more dB, for a total of 6 dB. If it doubled again to 40 watts, that would be a total of 9 dB.

If the signal increases by a factor of 10, that is 10 dB. For example, if a signal went from 1 watt to 10 watts, that is an increase of 10 dB. If it increased again to 100 watts, that is 10 more dB, for a total of 20 dB.

If the power goes up, the change in dB is a positive number. If the power goes down, the change in dB will be a negative number.

Those facts will allow you to answer all of the dB questions on the test.

Which decibel value most closely represents a power increase from 5 watts to 10 watts?
A. 2 dB
B. 3 dB
C. 5 dB

D. 10 dB

Which decibel value most closely represents a power decrease from 12 watts to 3 watts?
A. -1 dB
B. -3 dB
C. -6 dB
D. -9 dB

Which decibel value represents a power increase from 20 watts to 200 watts?
A. 10 dB
B. 12 dB
C. 18 dB
D. 28 dB

The device used to measure resistance is an **ohmmeter**. The device used to measure voltage is a **voltmeter**. The device used to measure current is an **ammeter**. These three devices are often included in one package, with a switch on the front to switch from one to the other. This combination is referred to as a **multitester**, or sometimes a "VOM". When measuring current, you need to put the ammeter in **series** with the circuit you are measuring.

When you measure voltage, you keep the circuit hooked up, and measure the voltage **across** the battery or resistor ("in **parallel**" with the thing you are measuring). This measures the voltage of the battery, or the voltage drop across the resistor:

To measure resistance, the resistor must be removed from the circuit, and you connect one lead of the ohmmeter to each wire of the resistor:

When using a multitester, you need to make sure that it is switched to the proper type of reading. If you measure resistance and the thing you are measuring is hooked up to current, you will damage the meter. You will also damage the meter if it is set to measure current (amps) and you hook it directly across the battery.

Which instrument would you use to measure electric potential?
A. An ammeter
B. A voltmeter
C. A wavemeter
D. An ohmmeter

How is a voltmeter connected to a component to measure applied voltage?
A. In series
B. In parallel
C. In quadrature
D. In phase

When configured to measure current, how is a multimeter connected to a component?
A. In series
B. In parallel
C. In quadrature
D. In phase

Which instrument is used to measure electric current?
A. An ohmmeter
B. A wavemeter
C. A voltmeter
D. An ammeter

Which of the following can damage a multimeter?
A. Attempting to measure resistance using the voltage setting
B. Failing to connect one of the probes to ground
C. Attempting to measure voltage when using the resistance setting
D. Not allowing it to warm up properly

Which of the following measurements are made using a multimeter?
A. Signal strength and noise

B. Impedance and reactance
C. Voltage and resistance
D. All these choices are correct

Which of the following precautions should be taken when measuring circuit resistance with an ohmmeter?
A. Ensure that the applied voltages are correct
B. Ensure that the circuit is not powered
C. Ensure that the circuit is grounded
D. Ensure that the circuit is operating at the correct frequency

When you become a ham, you should learn how to solder. It's not difficult, and a soldering iron and solder are very inexpensive.

For the test, you need to know that acid-core solder (the type used for plumbing) should never be used for electronics. You should also know that a "cold" (bad) solder connection has a rough or lumpy surface.

Which of the following types of solder should not be used for radio and electronic applications?

A. Acid-core solder
B. Lead-tin solder
C. Rosin-core solder
D. Tin-copper solder

What is the characteristic appearance of a cold tin-lead solder joint?
A. Dark black spots
B. A bright or shiny surface
C. A rough or lumpy surface
D. Excessive solder

In addition to resistors, there are some other electronic components that you should be able to identify for the test. You should also have a basic understanding of what they do.

An inductor is nothing more than a coil of wire. It stores energy in a magnetic field. Direct current can pass easily through an inductor, but alternating current has difficulty passing through one. The size of an inductor (in other words, its **inductance,** or ability to store energy in a magnetic field) is measured in Henries (or millihenries, more commonly). Here is the symbol for an inductor:

INDUCTOR

A variable inductor is simply a coil where the user can change the inductance by turning a knob or by some other means. This can be done either by physically connecting at a different spot, or sometimes by inserting a material that changes the inductance. Here is the symbol for a variable inductor:

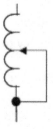

VARIABLE INDUCTOR

A capacitor, in its simplest form, is two metal plates separated by air. It is used to store energy in an electric field. Direct current is unable to pass through a capacitor, but alternating current is able to do so. The size of a capacitor (in other words, its capacitance, or ability to store energy in an electric field) is measured in Farads (or microfarads or picofarads, more commonly). Since a capacitor can store energy, if you hook an ohmmeter to a circuit containing a large capacitor, the meter might initially show a small resistance, but the resistance will gradually increase. For the test, you need to know that if an ohmmeter starts with a low resistance but slowly goes up, this means that there is a capacitor in the circuit. Here is the symbol for a capacitor:

CAPACITOR

An inductor and capacitor are usually used together to make a **tuned circuit**, also known as a **resonant circuit**. When you adjust the tuning dial on a radio, you are usually adjusting a variable capacitor. It is usually hooked to an inductor, and when you adjust the capacitance, the frequency changes. One of the questions refers to this as a kind of "filter," since it allows only one frequency to filter through.

A switch is used to connect or disconnect electric circuits. Here is the symbol for a switch:

SWITCH

A switch is sometimes described by the number of "poles" and "throws" it has. There are two kinds of switches mentioned on the test. A "single-pole single-throw" (SPST) switch, which is the type of switch shown above, and you need to recognize the symbol. The other kind of switch on the test is a single-pole double-throw (SPDT), which is used to switch a single circuit to two different circuits.

A relay is an electrically controlled switch. It is usually turned on and off by an electromagnet.

A fuse is used to protect other circuit components from too much current. A fuse consists of a small wire, usually inside of a glass tube. If there is too much current for the circuit, the fuse literally burns in half, and stops the flow of current. As one of the questions puts it, the fuse will "interrupt power in case of overload." You should never use a fuse larger than the intended size, because the excessive current could cause a fire. Here is the symbol for a fuse:

FUSE

A diode is a component that allows current to flow in one direction, but

DIODE

not the other. Here is the symbol for a diode:

Current will flow if the side of the diode with the arrow (the "cathode") is hooked up to the positive terminal of the source of electricity, and the straight line (the "anode") is hooked up to the negative terminal. Current will not flow if the diode is hooked up the other way. Even when current is flowing, there will be some voltage drop from a diode, and the amount of this forward voltage drop will vary from diode to diode.

There is usually a stripe printed on the diode, which is on the cathode side. This is easy to remember, because this is the same side of the diode that is the straight line in the schematic symbol.

Because a diode only allows current to flow in one direction, it is often used to convert alternating current (AC) into direct current (DC). When it is used for this purpose, it can be called a rectifier.

A special type of diode is the LED, or Light Emitting Diode. As its name implies, it gives off light whenever forward current is passing through it. An LED is commonly used as a visual indicator (such as a light showing that the power for a device is turned on). Here is the symbol for an LED:

LED (LIGHT EMITTING DIODE)

Here is the diagram for one type of transistor:

TRANSISTOR

The transistor shown here is called a "bipolar junction" transistor. For the test, you need to know that a bipolar junction transistor has three connections called the emitter, base, and collector. Most transistors are made of three regions of semiconductor material. Another common type of transistor is mentioned on the test, and that is the FET, or "Field Effect Transistor". The three connections on an FET are called the gate, source, and drain.

A transistor can be used to make an electronic switch or amplifier. One question says that a transistor can provide power gain. A small current or voltage is hooked to one side of the transistor, and this allows a smaller current or voltage to flow on the other side. The larger current is identical to the original current, but is much stronger. The term that describes how much a transistor can amplify a signal is the "gain".

In one diagram on the exam, the transistor is used to control the flow of current.

A transformer consists of two coils of wire that are located very close to one another. The main purpose of a transformer is to change the voltage of an alternating current.

A meter is simply any device that displays the strength of some quantity on a numeric scale. We've already talked about voltmeters, ammeters, and ohmmeters.

A voltage regulator (or simply "regulator") is a device that can be used to control amount of voltage from a power supply.

An integrated circuit (IC) is what is commonly known as a "chip". It combines several semiconductors and other devices into one small package.

What electrical component stores energy in an electric field?
A. Resistor
B. Capacitor
C. Inductor
D. Diode

What type of electrical component consists of conductive surfaces separated by an insulator?
A. Resistor
B. Potentiometer
C. Oscillator
D. Capacitor

What reading indicates that an ohmmeter is connected across a large, discharged capacitor?
A. Increasing resistance with time
B. Decreasing resistance with time
C. Steady full-scale reading
D. Alternating between open and short circuit

What type of electrical component stores energy in a magnetic field?
A. Resistor
B. Capacitor
C. Inductor
D. Diode

What electrical component is usually composed of a coil of wire?
A. Switch
B. Capacitor
C. Diode
D. Inductor

What electrical component is used to protect other circuit components from current overloads?
A. Fuse
B. Capacitor
C. Inductor
D. All of these choices are correct

What electronic component allows current to flow in only one direction?
A. Resistor
B. Fuse
C. Diode
D. Driven Element

Which is true about forward voltage drop in a diode?
A. It is lower in some diode types than in others
B. It is proportional to peak inverse voltage
C. It indicates that the diode is defective
D. It has no impact on the voltage delivered to the load

Which of these components can be used as an electronic switch?
A. Varistor
B. Potentiometer
C. Transistor
D. Thermistor

Which of the following components can consist of three regions of semiconductor material?
A. Alternator
B. Transistor
C. Triode
D. Pentagrid converter

Which of the following can provide power gain?
A. Transformer
B. Transistor
C. Reactor

D. Resistor

How is the cathode lead of a semiconductor diode usually identified?
A. With the word "cathode"
B. With a stripe
C. With the letter C
D. With the letter K

What causes a light-emitting diode (LED) to emit light?
A. Forward current
B. Reverse current
C. Capacitively-coupled RF signal
D. Inductively-coupled RF signal

What does the abbreviation FET stand for?
A. Frequency Emission Transmitter
B. Fast Electron Transistor
C. Free Electron Transmitter
D. Field Effect Transistor

What type of transistor has a gate, drain, and source?
A. Varistor
B. Field-effect
C. Tesla-effect
D. Bipolar junction

What are the names of the electrodes of a bipolar junction transistor?
A. Signal, bias, power
B. Emitter, base, collector
C. Input, output, supply
D. Pole one, pole two, output

What are the names of the electrodes of a diode?

A. Plus and minus
B. Source and drain
C. Anode and cathode
D. Gate and base

What is the term that describes a device's ability to amplify a signal?
A. Gain
B. Forward resistance
C. Forward voltage drop
D. On resistance

This diagram is used for the following questions:

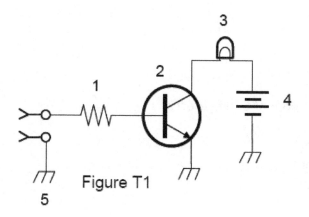

Figure T1

What is component 2 in figure T1?
A. Resistor
B. Transistor
C. Indicator lamp
D. Connector

What is the function of component 2 in Figure T1?
A. Give off light when current flows through it

B. Supply electrical energy
C. Control the flow of current
D. Convert electrical energy into radio waves

This diagram is used for the following questions:

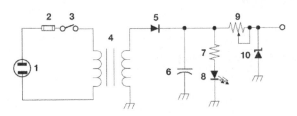

Figure T-2

What is component 6 in figure T2?
A. Resistor
B. Capacitor
C. Regulator IC
D. Transistor

What is component 8 in figure T2?
A. Resistor
B. Inductor
C. Regulator IC
D. Light emitting diode

What is component 4 in figure T2?
A. Variable inductor
B. Double-pole switch
C. Potentiometer
D. Transformer

Technician Study Guide

What type of switch is represented by component 3 in figure T2?
A. Single-pole single-throw
B. Single-pole double-throw
C. Double-pole single-throw
D. Double-pole double-throw

What is the function of an SPDT switch?
A. A single circuit is opened or closed
B. Two circuits are opened or closed
C. A single circuit is switched between one of two other circuits
D. Two circuits are each switched between one of two other circuits

This diagram is used for the following question:

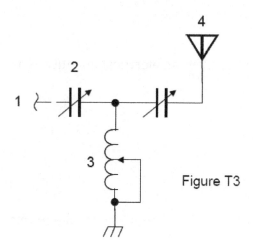

Figure T3

What is component 3 in figure T3?
A. Connector

B. Meter
C. Variable capacitor
D. Variable inductor

Which of the following devices or circuits changes an alternating current into a varying direct current signal?
A. Transformer
B. Rectifier
C. Amplifier
D. Reflector

What is a relay?
A. An electrically-controlled switch
B. A current controlled amplifier
C. An optical sensor
D. A pass transistor

Which of the following displays an electrical quantity as a numeric value?
A. Potentiometer
B. Transistor
C. Meter
D. Relay

What type of circuit controls the amount of voltage from a power supply?
A. Regulator
B. Oscillator
C. Filter
D. Phase inverter

What component changes 120 V AC power to a lower AC voltage for other uses?

A. Variable capacitor
B. Transformer
C. Transistor
D. Diode

Which of the following is commonly used as a visual indicator?
A. LED
B. FET
C. Zener diode
D. Bipolar transistor

What is the name of a device that combines several semiconductors and other components into one package?
A. Transducer
B. Multi-pole relay
C. Integrated circuit
D. Transformer

What is the purpose of a fuse in an electrical circuit?
A. To prevent power supply ripple from damaging a circuit
B. To interrupt power in case of overload
C. To limit current to prevent shocks
D. All of these choices are correct

Why should a 5-ampere fuse never be replaced with a 20-ampere fuse?
A. The larger fuse would be likely to blow because it is rated for higher current
B. The power supply ripple would greatly increase
C. Excessive current could cause a fire
D. All these choices are correct

What describes the ability to store energy in an electric field?
A. Inductance
B. Resistance
C. Tolerance
D. Capacitance

What is the basic unit of capacitance?
A. The farad
B. The ohm
C. The volt
D. The henry

What describes the ability to store energy in a magnetic field?
A. Admittance
B. Capacitance
C. Resistance
D. Inductance

What is the unit of inductance?
A. The coulomb
B. The farad
C. The henry
D. The ohm

Which of the following is combined with an inductor to make a tuned circuit?
A. Resistor
B. Zener diode
C. Potentiometer
D. Capacitor

Which of the following is a resonant or tuned circuit?
A. An inductor and a capacitor connected in series or parallel to form a filter
B. A linear voltage regulator
C. A resistor circuit used for reducing standing wave ratio
D. A circuit designed to provide high-fidelity audio

2 RADIO WAVES

As you probably already know, radio amateurs communicate using radio waves. The transmitter and antenna cause radio waves to be sent out in all directions. A radio wave is a type of "electromagnetic" wave. It has both an electrical and magnetic component. For one question on the test, you need to know that the electric field and the magnetic field are at right angles.

Radio waves have a frequency. When you adjust the tuning dial on a normal AM or FM radio, you are adjusting the frequency that you are receiving. Different stations are using different frequencies. The frequency of a radio wave, just like an alternating current, is measured in cycles per second, or **Hertz (Hz)**. Since radio waves have a frequency in the millions of cycles per second, it is most common to refer to their frequencies in kiloHertz (kHz) (thousand cycles per second) or MegaHertz (MHz)(million cycles per second).

Radio signals also have a **wavelength**, which is measured in **meters**. Signals with a lower frequency have a long wavelength. Signals with a high frequency have a short wavelength. The frequency times the wavelength equals the speed of light, which is 300 million meters per second.

If you know the frequency, then you can calculate the wavelength, and vice versa. When you make this calculation, it is easiest to work in Megahertz, since you won't have so many zeros to worry about. If you're working in MHz, here are the formulas:

Wavelength = 300 divided by frequency.

Frequency = 300 divided by wavelength.

So a frequency of 50 MHz would have a wavelength of 6 meters: 300 / 50 = 6.

If you know the wavelength is approximately 40 meters and want to know the frequency, you would use the formula 300 divided by 40, which would mean the frequency is approximately 7.5 MHz.

Amateurs are allowed to transmit on certain bands of frequencies, and these bands are named after the **approximate** wavelength. Therefore, the amateur band running from 50 - 54 MHz is called the "six meter band". The amateur band running from 7.0 - 7.3 MHz is called the "40 meter band". Note that these wavelengths are only approximate.

When you get your Technician class license, you will be allowed to transmit on the following frequencies, in the following bands.

80 Meters: 3.525-3.600 MHz: CW (Morse code) only, 200 watts maximum.

40 Meters: 7.025-7.125 MHz : CW only, 200 watts maximum.

15 Meters: 21.025-21.200 MHz: CW only, 200 watts maximum.

10 Meters: 28.000-28.300 MHz: CW, RTTY/Data--Maximum power 200 watts PEP, and 28.300-28.500 MHz: CW, Phone--Maximum power 200 watts PEP

6 Meters: 50.0-50.1 MHz: CW Only, and 50.1-54.0 MHz: All modes

2 Meters: 144.0-144.1 MHz: CW Only, and 144.1-148.0 MHz: All modes

1.25 Meters: 219-220 MHz (fixed digital message forwarding systems only), and 222.00-225.00 MHz: All modes

70 Centimeters: 420.0-450.0 MHz: All modes

33 Centimeters: 902.0-928.0 MHz: All modes

23 Centimeters: 1240-1300 MHz: All modes

There are also allocations on higher frequencies that are not covered on the test. In general, the maximum power allowed below 30 MHz is 200 watts, and above 30 MHz, 1500 watts.

When you get these questions on the test, even if you have forgotten the exact frequencies, you can often eliminate wrong answers by remembering the formulas for converting from frequency to wavelength.

The frequencies between 3-30 MHz are called "HF" or "High Frequency". The frequencies between 30-300 MHz are called "VHF" or "Very High Frequency". The frequencies between 300-3000 MHz are called "UHF" or "Ultra High Frequency".

Some radio signals (usually HF signals, between 3-30 MHz) can travel long distances around the earth because they are reflected by the a region of the atmosphere called the ionosphere.

When you set your transmitter frequency, it is very important to know how close to the band edge you can transmit. Your signal does not take up just the frequency shown on the dial. Your signal might be several kHz wide, depending on what mode you are using. Also, even the best transmitters may drift in frequency some, and the dial markings may not be completely accurate.

Also, all modes are not permitted on all frequencies. We'll discuss operating modes later on. But when you memorize the frequency bands you are allowed to use, remember which frequencies allow only a certain mode.

And keep in mind that some of our bands are shared with other services, both in the United States and other countries. For example, we share the 23 cm band with radiolocation stations outside the United States. If you learn that you are interfering with such a station, you must stop operating or take steps to eliminate the interference. This is particularly true if amateurs have only a **secondary** allocation. This means that there might be non-amateur stations in that band, and amateurs must avoid interfering with them. For the test, you need to know that the Radionavigation Service is always protected from interference from amateur signals.

There are some specific frequencies that are covered on the test. For example, the national FM simplex calling frequency on 2 meters is 146.520 MHz.

A repeater is a station, often set up by a club, that retransmits signals of other stations to increase their range. You transmit on one frequency, and receive on another frequency. On 2 meters, the difference in frequency (which is called the "offset" or "split") is usually 600 kHz. On the 70 cm band, the offset is usually 5 MHz. If you want to listen on the repeater's input frequency, you would use the "reverse split" function of the radio. If another station is close to you, but not close to the repeater, then you might be able to hear them directly, but not through the repeater. Therefore, you would use the reverse split function to listen on the input frequency.

On which HF bands does a Technician class operator have phone privileges?
A. None
B. 10 meter band only
C. 80 meter, 40 meter, 15 meter and 10 meter bands
D. 30 meter band only

Which of the following VHF/UHF frequency ranges are limited to CW only?
A. 50.0 MHz to 50.1 MHz and 144.0 MHz to 144.1 MHz
B. 219 MHz to 220 MHz and 420.0 MHz to 420.1 MHz
C. 902.0 MHz to 902.1 MHZ
D. All of these choices are correct

What is the maximum peak envelope power output for Technician class operators using their assigned portions of the HF bands?
A. 200 watts
B. 100 watts
C. 50 watts
D. 10 watts

Except for some specific restrictions, what is the maximum peak envelope power output for Technician class operators using frequencies above 30 MHz?
A. 50 watts
B. 100 watts
C. 500 watts
D. 1500 watts

Which of the following frequency ranges are available for phone operation by Technician licensees?
A. 28.050 MHz to 28.150 MHz
B. 28.100 MHz to 28.300 MHz
C. 28.300 MHz to 28.500 MHz
D. 28.500 MHz to 28.600 MHz

How are US amateurs restricted in segments of bands where the Amateur Radio Service is secondary?
A. U.S. amateurs may find non-amateur stations in those segments, and must avoid interfering with them
B. U.S. amateurs must give foreign amateur stations priority in those segments
C. International communications are not permitted in those segments
D. Digital transmissions are not permitted in those segments

What is the relationship between the electric and magnetic fields of an electromagnetic wave?

A. They travel at different speeds
B. They are in parallel
C. They revolve in opposite directions
D. They are at right angles

Which frequency is in the 6 meter band?
A. 49.00 MHz
B. 52.525 MHz
C. 28.50 MHz
D. 222.15 MHz

Which amateur band includes 146.52 MHz?
A. 6 meters
B. 20 meters
C. 70 centimeters
D. 2 meters

What are the two components of a radio wave?
A. Impedance and reactance
B. Voltage and current
C. Electric and magnetic fields
D. Ionizing and non-ionizing radiation

What is the velocity of a radio wave traveling through free space?
A. Speed of light
B. Speed of sound
C. Speed inversely proportional to its wavelength
D. Speed that increases as the frequency increases

What is the relationship between wavelength and frequency?
A. Wavelength gets longer as frequency increases
B. Wavelength gets shorter as frequency increases
C. Wavelength and frequency are unrelated
D. Wavelength and frequency increase as path length increases

What is the formula for converting frequency to approximate wavelength in meters?
A. Wavelength in meters equals frequency in hertz multiplied by 300
B. Wavelength in meters equals frequency in hertz divided by 300
C. Wavelength in meters equals frequency in megahertz divided by 300
D. Wavelength in meters equals 300 divided by frequency in megahertz

In addition to frequency, which of the following is used to identify amateur radio bands?
A. The approximate wavelength in meters
B. Traditional letter/number designators
C. Channel numbers
D. All these choices are correct

What frequency range is referred to as VHF?
A. 30 kHz to 300 kHz
B. 30 MHz to 300 MHz
C. 300 kHz to 3000 kHz
D. 300 MHz to 3000 MHz

What frequency range is referred to as UHF?
A. 30 to 300 kHz
B. 30 to 300 MHz
C. 300 to 3000 kHz
D. 300 to 3000 MHz

What frequency range is referred to as HF?
A. 300 to 3000 MHz
B. 30 to 300 MHz
C. 3 to 30 MHz
D. 300 to 3000 kHz

What is the approximate velocity of a radio wave in free space?

A. 150,000 meters per second
B. 300,000,000 meters per second
C. 300,000,000 miles per hour
D. 150,000 miles per hour

Which region of the atmosphere can refract or bend HF and VHF radio waves?
A. The stratosphere
B. The troposphere
C. The ionosphere
D. The mesosphere

Why should you not set your transmit frequency to be exactly at the edge of an amateur band or sub-band?
A. To allow for calibration error in the transmitter frequency display
B. So that modulation sidebands do not extend beyond the band edge
C. To allow for transmitter frequency drift
D. All of these choices are correct

What is the national calling frequency for FM simplex operations in the 2 meter band?
A. 146.520 MHz
B. 145.000 MHz
C. 432.100 MHz
D. 446.000 MHz

Which of the following is a common repeater frequency offset in the 2 meter band?
A. Plus or minus 5 MHz
B. Plus or minus 600 kHz
C. Plus or minus 500 kHz
D. Plus or minus 1 MHz

What is a common repeater frequency offset in the 70 cm band?
A. Plus or minus 5 MHz

B. Plus or minus 600 kHz
C. Plus or minus 500 kHz
D. Plus or minus 1 MHz

What is meant by "repeater offset?"
A. The difference between a repeater's transmit frequency and its receive frequency
B. The repeater has a time delay to prevent interference
C. The repeater station identification is done on a separate frequency
D. The number of simultaneous transmit frequencies used by a repeater

How is a VHF/UHF transceiver's "reverse" function used?
A. To reduce power output
B. To increase power output
C. To listen on a repeater's input frequency
D. To listen on a repeater's output frequency

How may amateurs use the 219 to 220 MHz segment of 1.25 meter band?
A. Spread spectrum only
B. Fast-scan television only
C. Emergency traffic only
D. Fixed digital message forwarding systems only

3 ANTENNAS

We'll talk about transmitters later. But as you know, a transmitter is a device that generates radio signals. The signal that comes out of the radio is known as "**RF**", or a **Radio Frequency** signal. In order to change into a radio wave, it needs to go to an antenna. Similarly, to receive a signal, you need some sort of antenna. The antenna changes the radio waves into an RF signal, which the receiver will be able to receive.

Here is the schematic symbol for an antenna:

ANTENNA

Two common antennas are the quarter wave vertical antenna and the half wave dipole antenna. For the test, you will need to know the approximate length of these antennas. However, the test practically gives you the answer, because it will ask for the length of a half-wave dipole for six meters, or for the length of a quarter wave vertical for two meters.

If you forget the answer, it is easy to calculate this length in your head. As the name implies, a half wave dipole is approximately half as long as the wavelength it is designed for. So for six meters, this antenna would be 1/2 of 6, or about 3 meters long. One meter is about 3 feet, so the antenna would be about 9 feet long, which is 108 inches. The closest answer is 112 inches, and this is the correct answer.

Another question is the length of a quarter wave vertical for two meters. Again, as the name implies, the length is 1/4 the wavelength, or 1/4 of 2 meters, which is 1/2 meter, or about 18 inches. The closest answer on the test is 19 inches, and this is the correct answer.

Another type of vertical antenna that is often used on two meters is the 5/8 wave antenna. This antenna offers a lower angle of radiation and more gain than a 1/4 wavelength antenna and usually provides improved coverage. Mobile antennas are often mounted in the center of the vehicle's roof, because this position normally provides the most uniform radiation pattern. One way to electrically "lengthen" a short antenna is to "load" it by inserting an inductor at the bottom.

A half wave dipole antenna is normally installed horizontally (in other words, parallel to the earth's surface). The signal that comes from a horizontal antenna is "**horizontally polarized**", and for making local contacts, works best when communicating with other stations with horizontally polarized antennas. For one question on the test, you need to know that the polarization of a radio wave is defined by the orientation of the electric field.

As noted above, antennas are tuned to a particular frequency, based on their length. To change the frequency of an antenna, you would lengthen it to work on a lower frequency, or make it shorter to work on a higher frequency. This is easy to remember if you think in terms of the wavelength. A six meter antenna (lower frequency) will be longer than a two meter antenna (higher frequency).

A dipole antenna transmits and receives best broadside to the antenna. In other words, if it is running north to south, it will "get out" best to the east and west.

As mentioned above, a vertical antenna is generally one quarter wavelength long. The signal is vertically polarized, and for local work, works best with a vertical antenna at the station you are contacting. For one question on the test, you need to know that the electric field is perpendicular to the Earth.

One question lists a number of antennas and asks which has the greatest gain. The answer is the Yagi antenna.

The gain of a directional antenna means how much stronger the signal is in one direction, as compared to other directions.

The antenna that comes with most handheld transceivers is called a "rubber duck" antenna. It does not transmit or receive as efficiently as a full size antenna. In particular, they won't work very well inside a car, because the signals will be weaker.

The cable that runs between the radio and the antenna is called the feed line. The most common kind of feed line is coaxial cable. Coaxial cable has a center conductor, and an outer shield. The most commonly used type of coax by amateurs is called "50 ohm" coax. Coaxial cable is the most common because it is easy to use and has few special installation considerations. However, coax can have more loss than other types of feed line. As the frequency increases, the signal loss also increases. Some sources of loss in coaxial cable are water intrusion, high SWR, and multiple connectors in the line.

The connectors on coaxial cable can be of various types. The two most common types are covered on the test. The PL-259 connector is most commonly used on HF and VHF frequencies. The type N connector is most suitable for UHF frequencies.

When used outdoors, the coax connectors should be sealed against water, because the water will cause the loss to increase. The most common reason for failure of coaxial cable is moisture contamination. In some cases, ultraviolet light from the sun can damage the outer insulator of the cable and allow water to enter. In particular, you must be very careful with "air core" coaxial cable, because it requires special techniques to prevent water absorption. However, air insulated coax has the lowest loss on VHF and UHF frequencies.

The two types of cable mentioned on the test are types RG-58 and RG-213. Type RG-213 is thicker, and has lower loss.

Another type of feed line mentioned on the test is air-insulated hard line. This type of cable has the lowest loss on VHF and UHF.

When power is lost in a feed line, this results in less signal making it to your antenna. The power that is lost actually causes the feed line to heat up.

The antenna should be matched to the feed line, and to the transmitter. Whether or not these are properly matched is usually done with a device called an SWR (standing wave ratio) meter. You want to adjust things so that the SWR is as low as possible. This is important because it allows for the efficient transfer of power and reduces losses. The SWR is a measure of how well the radio and feed line are matched to the load (the antenna). The SWR meter is hooked in series with the feed line, between the transmitter and antenna. (An RF power meter is also installed between the transmitter and antenna.) When selecting an SWR meter, you need to keep in mind the power level and frequencies that you'll be using it at.

The SWR should be as low as possible. The lowest possible reading is 1:1 (when spoken, "1 to 1"). Most modern transmitters begin to automatically reduce their power output when the SWR reaches about

2:1. A high SWR reading (for example 4:1) means that there is an impedance mismatch.

One way to reduce the SWR is to use an antenna tuner. This device matches the antenna system impedance to the transceiver's output impedance. If the SWR changes erratically, this could be the result of a loose connection in an antenna or feed line.

Another device that can be used in place of an SWR meter would be a directional wattmeter.

One question asks about antenna **loading**. "Loading" just means that you attach an inductor (coil) to the antenna to make it appear longer than it really is.

Which of the following should be considered when selecting an accessory SWR meter?
A. The frequency and power level at which the measurements will be made
B. The distance that the meter will be located from the antenna
C. The types of modulation being used at the station
D. All these choices are correct

Where should an RF power meter be installed?
A. In the feed line, between the transmitter and antenna
B. At the power supply output
C. In parallel with the push-to-talk line and the antenna
D. In the power supply cable, as close as possible to the radio

What does the abbreviation "RF" refer to?
A. Radio frequency signals of all types
B. The resonant frequency of a tuned circuit
C. The real frequency transmitted as opposed to the apparent frequency
D. Reflective force in antenna transmission lines

This diagram is used in the following question:

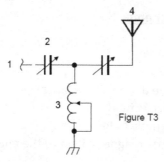

Figure T3

What is component 4 in figure T3?
A. Antenna
B. Transmitter
C. Dummy load
D. Ground

What is the approximate length, in inches, of a quarter-wavelength vertical antenna for 146 MHz?
A. 112
B. 50
C. 19
D. 12

What is the approximate length, in inches, of a 6 meter 1/2-wavelength wire dipole antenna?
A. 6
B. 50
C. 112
D. 236

What is a beam antenna?
A. An antenna built from aluminum I-beams
B. An omnidirectional antenna invented by Clarence Beam
C. An antenna that concentrates signals in one direction
D. An antenna that reverses the phase of received signals

Which of the following describes a simple dipole oriented parallel to Earth's surface?
A. A ground-wave antenna
B. A horizontally polarized antenna
C. A travelling-wave antenna
D. A vertically polarized antenna

What property of a radio wave defines its polarization?
A. The orientation of the electric field
B. The orientation of the magnetic field
C. The ratio of the energy in the magnetic field to the energy in the electric field
D. The ratio of the velocity to the wavelength

What is a disadvantage of the short, flexible antenna supplied with most handheld radio transceivers, compared to a full-sized quarter-wave antenna?
A. It has low efficiency
B. It transmits only circularly polarized signals
C. It is mechanically fragile
D. All these choices are correct

Which of the following increases the resonant frequency of a dipole antenna?
A. Lengthening it
B. Inserting coils in series with radiating wires
C. Shortening it
D. Adding capacitive loading to the ends of the radiating wires

Which of the following types of antenna offers the greatest gain?
A. 5/8 wave vertical
B. Isotropic
C. J pole
D. Yagi

What is a disadvantage of using a handheld VHF transceiver with a flexible antenna inside a vehicle?
A. Signal strength is reduced due to the shielding effect of the vehicle
B. The bandwidth of the antenna will decrease, increasing SWR
C. The SWR might decrease, decreasing the signal strength
D. All these choices are correct

In which direction does a half-wave dipole antenna radiate the strongest signal?
A. Equally in all directions
B. Off the ends of the antenna
C. In the direction of the feed line
D. Broadside to the antenna

What is antenna gain?
A. The additional power that is added to the transmitter power
B. The additional power that is required in the antenna when transmitting on a higher frequency
C. The increase in signal strength in a specified direction compared to a reference antenna
D. The increase in impedance on receive or transmit compared to a reference antenna

What is an advantage of a 5/8 wavelength whip antenna for VHF or UHF mobile service?
A. It has more gain than a 1/4-wavelength antenna
B. It radiates at a very high angle
C. It eliminates distortion caused by reflected signals
D. It has 10 times the power gain of a 1/4 wavelength whip

What happens to power lost in a feed line?
A. It increases the SWR
B. It is radiated as harmonics
C. It is converted into heat
D. It distorts the signal

What is the most common impedance of coaxial cables used in amateur radio?

A. 8 ohms
B. 50 ohms
C. 600 ohms
D. 12 ohms

Why is coaxial cable the most common feed line for amateur radio antenna systems?
A. It is easy to use and requires few special installation considerations
B. It has less loss than any other type of feed line
C. It can handle more power than any other type of feed line
D. It is less expensive than any other type of feed line

What happens as the frequency of a signal in coaxial cable is increased?
A. The characteristic impedance decreases
B. The loss decreases
C. The characteristic impedance increases
D. The loss increases

Which of the following RF connector types is most suitable for frequencies above 400 MHz?
A. UHF (PL-259/SO-239)
B. Type N
C. RS-213
D. DB-25

Which of the following is true of PL-259 type coax connectors?
A. They are preferred for microwave operation
B. They are watertight
C. They are commonly used at HF and VHF frequencies
D. They are a bayonet-type connector

What is the electrical difference between RG-58 and RG-213 coaxial cable?
A. There is no significant difference between the two types
B. RG-58 cable has two shields
C. RG-213 cable has less loss at a given frequency
D. RG-58 cable can handle higher power levels

Which of the following types of feed line has the lowest loss at VHF and UHF?
A. 50-ohm flexible coax
B. Multi-conductor unbalanced cable
C. Air-insulated hard line
D. 75-ohm flexible coax

What is a benefit of low SWR?
A. Reduced television interference
B. Reduced signal loss
C. Less antenna wear
D. All these choices are correct

What is the major function of an antenna tuner (antenna coupler)?
A. It matches the antenna system impedance to the transceiver's output impedance
B. It helps a receiver automatically tune in weak stations
C. It allows an antenna to be used on both transmit and receive
D. It automatically selects the proper antenna for the frequency band being used

What can cause erratic changes in SWR?
A. Local thunderstorm
B. Loose connection in the antenna or feed line
C. Over-modulation
D. Overload from a strong local station

What is standing wave ratio (SWR)?
A. A measure of how well a load is matched to a transmission line
B. The ratio of amplifier power output to input
C. The transmitter efficiency ratio
D. An indication of the quality of your station's ground connection

What reading on an SWR meter indicates a perfect impedance match between the antenna and the feed line?
A. 50:50
B. Zero
C. 1:1
D. Full Scale

What does an SWR reading of 4:1 indicate?
A. Loss of -4dB
B. Good impedance match
C. Gain of +4dB
D. Impedance mismatch

Which instrument can be used to determine SWR?
A. Voltmeter
B. Ohmmeter
C. Iambic pentameter
D. Directional wattmeter

Which of the following causes failure of coaxial cables?
A. Moisture contamination
B. Solder flux contamination
C. Rapid fluctuation in transmitter output power
D. Operation at 100% duty cycle for an extended period

Why should the outer jacket of coaxial cable be resistant to ultraviolet light?
A. Ultraviolet resistant jackets prevent harmonic radiation
B. Ultraviolet light can increase losses in the cable's jacket
C. Ultraviolet and RF signals can mix together, causing interference
D. Ultraviolet light can damage the jacket and allow water to enter the cable

What is a disadvantage of air core coaxial cable when compared to foam or solid dielectric types?
A. It has more loss per foot
B. It cannot be used for VHF or UHF antennas
C. It requires special techniques to prevent moisture in the cable
D. It cannot be used at below freezing temperatures

Which of the following describes a type of antenna loading?
A. Electrically lengthening by inserting inductors in radiating elements
B. Inserting a resistor in the radiating portion of the antenna to make it resonant
C. Installing a spring in the base of a mobile vertical antenna to make it more flexible
D. Strengthening the radiating elements of a beam antenna to better resist wind damage

Which of the following is a source of loss in coaxial feed line?
A. Water intrusion into coaxial connectors
B. High SWR
C. Multiple connectors in the line
D. All these choices are correct

4 PROPAGATION OF RADIO WAVES

As we discussed previously, radio waves are essentially the same as light waves, but have a different frequency and wavelength. And in general, they behave the same as light waves. You can usually see things only if there is nothing between you and the thing you want to see. And that is the same, in general, for radio waves. In particular (especially on VHF and UHF frequencies), you can communicate by radio if you, while standing at one antenna, are able to see the other antenna.

This is why, especially on VHF and UHF, it is important for the antenna to be at a high location. Your antenna can "see" further if it is higher up, just as you can see further if you in a tall building, than you would be able to on the ground. You cannot see over the horizon. And in general, your antenna cannot see over the horizon. In radio, this is called the **radio horizon**. The radio horizon is the distance at which radio signals between two points are effectively blocked by the curvature of the Earth.

Unlike light waves, radio signals have some ability to penetrate solid objects. (For example, this is why UHF signals generally work better from inside buildings than VHF signals—UHF signals have a greater ability to penetrate solid object such as buildings.) But this ability is limited by something as solid as the earth itself. So the radio horizon is somewhat further away than the normal horizon, generally about 10% further away. This is because the atmosphere has a limited ability to refract radio waves. But if the other station is over the horizon, you won't be able to contact them, unless there is some other type of propagation available.

Like light waves, radio waves have some ability to be reflected. For example, if you need to communicate with a repeater a long distance away, but your antenna is not high enough, it might be possible to bounce your signal off a large object such as a building or water tower. If you have a directional antenna, instead of pointing it directly toward the repeater, you can try pointing it at a nearby building.

But the most common object that amateurs use to reflect their signals great distances is the ionosphere, a very high layer of the atmosphere. The ionosphere is like a large mirror above the earth. If the ionosphere reflected light (which it does not), then if someone shone a flashlight at it from a very long distance, you would be able to see the reflection.

Fortunately, the ionosphere does reflect radio waves in this manner. The ionosphere generally reflects radio waves in the HF portion of the spectrum (approximately 3-30 MHz). This type of propagation is commonly called **skip**, because the signal "skips" off the ionosphere. It normally does not reflect VHF and UHF signals. This is why VHF and UHF signals can only be heard "line of sight", that is, they can't be heard over the radio horizon. Because the ionosphere is not a perfect mirror, signals may come from all directions. They may interfere with each other to some extent, which can cause signals to fade in and out somewhat, even during times of good propagation.

Generally, skip propagation through the ionosphere is more common on higher frequencies during the day, and lower frequencies at night. For example, skip propagation on the 10 meter band is most common during the day. Propagation is also affected by sunspots. In particular, you should know for the test that the ten and six meter bands can be used for long-distance communications during the peak of the sunspot cycle.

On HF frequencies, skip signals are usually reflected by the highest layer of the ionosphere, the F layer. Sometimes, VHF signals are reflected by the next lower layer, the E layer. Since this type of reflection is somewhat uncommon and unpredictable, it is called **sporadic E**. Sporadic E is most common on the 10, 6, and 2 meter bands.

As we mentioned previously, having the same orientation ("**polarization**") of the antennas (horizontal or vertical) is important for local contacts. It is less important for skip propagation, because the polarization of the signals becomes randomized while the signal goes through the ionosphere. Therefore, for "skip" signals, either horizontal or vertical antennas can be used. The polarization of a signal refers to the orientation of the electrical field of the signal.

Speaking of polarization, vertical antennas are normally used on VHF and UHF for FM transmissions. Horizontal antennas are normally used for SSB and CW transmissions. (We will explain these different types of transmissions later.) If you do not use the same polarization as the station you are contacting (in other words, you are using a vertical antenna and the other station is using a horizontal antenna), the signals will be significantly weaker.

On VHF and UHF, reflections from various objects may cause your signals to cancel out. This is called **multipath**, and it is sometimes cured by moving just a few feet. If a mobile station is moving while transmitting, this effect can be very pronounced, and can cause a rapid fluttering sound. This effect is called **picket fencing**. Multipath propagation is likely to cause error rates to increase with data transmission.

There are a few ways that can cause VHF and UHF signals to be propagated over the horizon. The most common is tropospheric ducting, which is responsible for over-the-horizon communication up to about 300 miles. This can be caused by temperature inversions in the atmosphere.

For UHF and microwave signals, vegetation can absorb radio waves and make these frequencies less useful. Fog and rain can affect some radio signals, but there is a trick question on the test. The test asks whether fog and light rain affect signals on six and ten meters. There is little effect on these bands, although there might be an effect on higher frequencies. On microwave frequencies, these can block your signal. So for the question of what can decrease your microwave range, the correct answer is precipitation.

There are other ways that VHF and UHF signals can sometimes be propagated over the horizon. For example, during periods of aurora, distant signals can be heard by auroral reflection. These signals exhibit rapid fluctuations in strength, and often sound distorted.

Knife-edge propagation is the phenomenon when signals are refracted around solid objects with sharp edges, such as a mountain range.

When meteors enter the atmosphere, signals can be reflected from them. Six meters is the best band for meteor scatter communication.

Why do VHF signal strengths sometimes vary greatly when the antenna is moved only a few feet?
A. The signal path encounters different concentrations of water vapor
B. VHF ionospheric propagation is very sensitive to path length
C. Multipath propagation cancels or reinforces signals
D. All these choices are correct

What effect does multi-path propagation have on data transmissions?
A. Transmission rates must be increased by a factor equal to the number of separate paths observed
B. Transmission rates must be decreased by a factor equal to the number of separate paths observed

C. No significant changes will occur if the signals are transmitted using FM
D. Error rates are likely to increase

What is the effect of vegetation on UHF and microwave signals?
A. Knife-edge diffraction
B. Absorption
C. Amplification
D. Polarization rotation

What is the effect of fog and rain on signals in the 10 meter and 6 meter bands?
A. Absorption
B. There is little effect
C. Deflection
D. Range increase

What weather condition might decrease range at microwave frequencies?
A. High winds
B. Low barometric pressure
C. Precipitation
D. Colder temperatures

What antenna polarization is normally used for long-distance CW and SSB contacts on the VHF and UHF bands?
A. Right-hand circular
B. Left-hand circular
C. Horizontal
D. Vertical

What happens when antennas at opposite ends of a VHF or UHF line of sight radio link are not using the same polarization?
A. The modulation sidebands might become inverted
B. Received signal strength is reduced
C. Signals have an echo effect
D. Nothing significant will happen

When using a directional antenna, how might your station be able to communicate with a distant repeater if buildings or obstructions are blocking the direct line of sight path?
A. Change from vertical to horizontal polarization
B. Try to find a path that reflects signals to the repeater
C. Try the long path
D. Increase the antenna SWR

What is the meaning of the term "picket fencing"?
A. Alternating transmissions during a net operation
B. Rapid flutter on mobile signals due to multipath propagation
C. A type of ground system used with vertical antennas
D. Local vs long-distance communications

What is a likely cause of irregular fading of signals propagated by the ionosphere?
A. Frequency shift due to Faraday rotation
B. Interference from thunderstorms
C. Intermodulation distortion
D. Random combining of signals arriving via different paths

Which of the following results from the fact that signals propagated by the ionosphere are elliptically polarized?
A. Digital modes are unusable
B. Either vertically or horizontally polarized antennas may be used for transmission or reception
C. FM voice is unusable
D. Both the transmitting and receiving antennas must be of the same polarization

Why are simplex UHF signals rarely heard beyond their radio horizon?
A. They are too weak to go very far
B. FCC regulations prohibit them from going more than 50 miles
C. UHF signals are usually not propagated by the ionosphere

D. UHF signals are absorbed by the ionospheric D region

What is a characteristic of VHF signals received via auroral backscatter?
A. They are often received from 10,000 miles or more
B. They are distorted and signal strength varies considerably
C. They occur only during winter nighttime hours
D. They are generally strongest when your antenna is aimed west

Which of the following types of propagation is most commonly associated with occasional strong signals on the 10, 6, and 2 meter bands from beyond the radio horizon?
A. Backscatter
B. Sporadic E
C. D region absorption
D. Gray-line propagation

Which of the following effects may allow radio signals to travel beyond obstructions between the transmitting and receiving stations?
A. Knife-edge diffraction
B. Faraday rotation
C. Quantum tunneling
D. Doppler shift

What mode is responsible for allowing over-the-horizon VHF and UHF communications to ranges of approximately 300 miles on a regular basis?
A. Tropospheric ducting
B. D layer refraction
C. F2 layer refraction
D. Faraday rotation

Why is the radio horizon for VHF and UHF signals more distant than the visual horizon?
A. Radio signals move somewhat faster than the speed of light
B. Radio waves are not blocked by dust particles
C. The atmosphere refracts radio waves slightly
D. Radio waves are blocked by dust particles

What band is best suited for communicating via meteor scatter?
A. 33 centimeters
B. 6 meters
C. 2 meters
D. 70 cm

What causes tropospheric ducting?
A. Discharges of lightning during electrical storms
B. Sunspots and solar flares
C. Updrafts from hurricanes and tornadoes
D. Temperature inversions in the atmosphere

What is generally the best time for long-distance 10 meter band propagation via the F region?
A. From dawn to shortly after sunset during periods of high sunspot activity
B. From shortly after sunset to dawn during periods of high sunspot activity
C. From dawn to shortly after sunset during periods of low sunspot activity
D. From shortly after sunset to dawn during periods of low sunspot activity

What is a characteristic of HF communication compared with communications on VHF and higher frequencies?
A. HF antennas are generally smaller
B. HF accommodates wider bandwidth signals
C. Long-distance ionospheric propagation is far more common on HF
D. There is less atmospheric interference (static) on HF

Which of the following bands may provide long-distance communications via the ionosphere's F region during the peak of the sunspot cycle?
A. 6 and 10 meters
B. 23 centimeters
C. 70 centimeters and 1.25 meters

D. All these choices are correct

5 RADIO EQUIPMENT AND SETTING UP A STATION

As you probably know, a transmitter is a piece of equipment that generates RF (radio) signals. If it is connected to an antenna, then the antenna will radiate radio waves. The receiver is the device that connects to the antenna that converts those RF signals into a form that you can hear. The process in a transmitter that combines speech with an RF carrier signal is called **modulation**.

A transceiver is simply a transmitter and receiver that are combined into one package. Normally, when using phone (voice) transmissions, it is switched from transmit to receive by pushing the push-to-talk (PTT) switch on the microphone. When the push-to-talk switch is pushed, this causes another switch internal to the transceiver to be switched. The PTT connection on the radio usually switches to transmit when grounded.

Most modern transceivers have two ways of entering the desired frequency. There is usually a VFO (variable frequency oscillator) control to manually set the frequency. In addition, to enable quick access to favorite frequencies, you can store the frequency in a memory channel. Frequencies can often be entered numerically with a keypad.

When using phone (voice) transmissions, the transmitter, of course, combines the radio signal with the speech signal. The part of the transmitter that does this is called the modulator.

FM transmitters sometimes used sub-audible tones to turn on the receiver ("open the squelch") at the other station, or a repeater. These are commonly known as "PL tones", but that term is a trademark, and is not used on the test. The correct term is CTCSS tones. If you can hear a repeater but are unable to access it, there might be a number of possible reasons. For access, the repeater might require either an audio tone burst, or a CTCSS tone, or a DCS (digital coded squelch). The problem might also be that you are using the wrong repeater offset.

The part of any transmitter (including a handheld transceiver, which is the one mentioned on the test) that increases the power output is the RF power amplifier. You need to know that most transmitters automatically reduce power as SWR increases to protect the output amplifier transistors. Just remember that if the transmitter's power goes down automatically, this is to protect the transmitter. If you remember that, you'll be able to pick out the right answer.

Microphone connections on transceivers generally are **not** standardized, and when wiring them up, keep in mind that there is usually a connection for the push-to-talk switch, and there is sometimes a connection to supply voltage to the microphone.

One control on most modern receivers and transceivers is the squelch function. The purpose of this control is to mute the output to the speaker when no signal is being received. This is a type of **carrier squelch**, since the presence of the carrier (radio signal) turns the speaker on. When you are trying to receive a weak FM signal, the squelch should be turned down—adjusted so that the receiver audio is on all the time.

Most mobile transceivers operate are designed to operate from 13.8 volts. Most cars have a 12 volt battery, but with the engine running, the alternator usually provides about 13.8 volts. For one question, you need to know that a 50 watt mobile radio would require a power supply of 13.8 volts at 12 amps. You also need to know that the negative power connection should usually be made at the 12 volt battery's chassis ground. You should use a regulated power supply to keep voltage fluctuations from damaging sensitive circuits in the radio. When hooking up the power supply, be sure to use short heavy wire to prevent the voltage from dropping too low.

To operate a radio in a vehicle, the best place to make the negative connection for the power is directly at the vehicles battery's chassis ground.

Radio transmitters can sometimes generate harmonics, which is a signal on an exact multiple of the original frequency. For example, if you are transmitting on 7 MHz, there could be harmonics on 14 MHz, 21 MHz, 28 MHz, etc. These signals could cause interference. A filter to reduce these signals can be placed between the transmitter and antenna. Other things that can cause interference include fundamental overload and spurious emissions.

An AM or FM broadcast radio might unintentionally receive an amateur transmission because the receiver is unable to reject strong signals from other bands. This kind of interference can often be eliminated with a filter at the antenna input of the affected receiver. Interference to cable TV can often be cured by making sure all TV coaxial connectors are installed properly

Radio frequency interference can also be caused by spurious emissions. There is one question on the test that asks which of the following can cause interference: fundamental overload, harmonics, or spurious emissions. The correct answer is that all three are correct. One question asks what you can do if your VHF transceiver is being overloaded by a nearby FM broadcast station. The correct answer is using a "band-reject filter," because you want your radio to reject signals on that one frequency.

If you are ever told that you are causing interference, the first thing you should do is to make sure that your station is working properly. One way to do this is to see whether you are causing interference to your own television. In resolving interference problems, all of the following items might be useful, depending on the situation: snap-on ferrite chokes, low-pass and high-pass filters, and band-reject and band-pass filters.

The **mixer** is the part of a receiver that causes the signal to change from one frequency to another. Many pieces of equipment, including receivers, contain a circuit that generates a signal of a specific frequency. This circuit is called and **oscillator.** The part of the receiver that converts the RF (radio) signal into an audio signal is called the detector or demodulator.

One of the most important attributes of a receiver is its **selectivity**. This is the receiver's ability to discriminate between multiple signals. **Sensitivity** is a receiver's ability to detect the presence of a signal.

Sometimes, in order to amplify weak signals, an RF preamplifier can be used with a receiver. The RF preamplifier is installed between the antenna and the receiver.

Some receivers or transceivers have a scanning function to allow you check a large range of frequencies. If you get the question about a radio with a "scanning function," remember that the right answer is the only one that has the word "scan"!

A transverter is a device that changes the output of a transceiver for one band to cover another band. In that case, the original transceiver is sometimes called the exciter. For example, a transceiver could convert a 28 MHz transceiver into a 222 MHz transceiver.

One important part of any station is a dummy load. A dummy load takes the place of the antenna, and allows you to test your transmitter without actually transmitting on the air. It prevents the radiation of signals while making tests. It consists of a non-inductive resistor and a heat sink.

What term describes the use of a sub-audible tone transmitted along with normal voice audio to open the squelch of a receiver?
A. Carrier squelch
B. Tone burst
C. DTMF
D. CTCSS

Which of the following could be the reason you are unable to access a repeater whose output you can hear?
A. Improper transceiver offset
B. You are using the wrong CTCSS tone
C. You are using the wrong DCS code
D. All these choices are correct

Which of the following can be used to enter a transceiver's operating frequency?
A. The keypad or VFO knob
B. The CTCSS or DTMF encoder
C. The Automatic Frequency Control
D. All these choices are correct

Which term describes the ability of a receiver to detect the presence of a signal?
A. Linearity
B. Sensitivity
C. Selectivity
D. Total Harmonic Distortion

Why are short, heavy-gauge wires used for a transceiver's DC power connection?

A. To minimize voltage drop when transmitting
B. To provide a good counterpoise for the antenna
C. To avoid RF interference
D. All these choices are correct

Why do most solid-state transmitters reduce output power as SWR increases beyond a certain level?
A. To protect the output amplifier transistors
B. To comply with FCC rules on spectral purity
C. Because power supplies cannot supply enough current at high SWR
D. To lower the SWR on the transmission line

What is a transceiver?
A. A device that combines a receiver and transmitter
B. A device for matching feed line impedance to 50 ohms
C. A device for automatically sending and decoding Morse code
D. A device for converting receiver and transmitter frequencies to another band

Which of the following is used to convert a signal from one frequency to another?
A. Phase splitter
B. Mixer
C. Inverter
D. Amplifier

What is the name of a circuit that generates a signal at a specific frequency?
A. Reactance modulator
B. Phase modulator
C. Low-pass filter
D. Oscillator

What device converts the RF input and output of a transceiver to another band?
A. High-pass filter
B. Low-pass filter
C. Transverter

D. Phase converter

What is the function of a transceiver's PTT input?
A. Input for a key used to send CW
B. Switches transceiver from receive to transmit when grounded
C. Provides a transmit tuning tone when grounded
D. Input for a preamplifier tuning tone

Which of the following describes combining speech with an RF carrier signal?-
A. Impedance matching
B. Oscillation
C. Modulation
D. Low-pass filtering

What is the purpose of a squelch function?
A. Reduce a CW transmitter's key clicks
B. Mute the receiver audio when a signal is not present
C. Eliminate parasitic oscillations in an RF amplifier
D. Reduce interference from impulse noise

How is squelch adjusted so that a weak FM signal can be heard?
A. Set the squelch threshold so that receiver output audio is on all the time
B. Turn up the audio level until it overcomes the squelch threshold
C. Turn on the anti-squelch function
D. Enable squelch enhancement

Which of the following is an appropriate power supply rating for a typical 50 watt output mobile FM transceiver?
A. 24.0 volts at 4 amperes
B. 13.8 volts at 4 amperes
C. 24.0 volts at 12 amperes
D. 13.8 volts at 12 amperes

Where should the negative power return of a mobile transceiver be connected in a vehicle?

A. At the 12 volt battery chassis ground
B. At the antenna mount
C. To any metal part of the vehicle
D. Through the transceiver's mounting bracket

What is a way to enable quick access to a favorite frequency or channel on your transceiver?
A. Enable the frequency offset
B. Store it in a memory channel
C. Enable the VOX
D. Use the scan mode to select the desired frequency

Which of the following can reduce overload of a VHF transceiver by a nearby commercial FM station?
A. Installing an RF preamplifier
B. Using double-shielded coaxial cable
C. Installing bypass capacitors on the microphone cable
D. Installing a band-reject filter

What does the scanning function of an FM transceiver do?
A. Checks incoming signal deviation
B. Prevents interference to nearby repeaters
C. Tunes through a range of frequencies to check for activity
D. Checks for messages left on a digital bulletin board

What device increases the transmitted output power from a transceiver?
A. A voltage divider
B. An RF power amplifier
C. An impedance network
D. All these choices are correct

Which term describes the ability of a receiver to discriminate between multiple signals?
A. Discrimination ratio

B. Sensitivity
C. Selectivity
D. Harmonic Distortion

Where is an RF preamplifier installed?
A. Between the antenna and receiver
B. At the output of the transmitter's power amplifier
C. Between a transmitter and antenna tuner
D. At the receiver's audio output

Which of the following can cause radio frequency interference?
A. Fundamental overload
B. Harmonics
C. Spurious emissions
D. All these choices are correct

What would cause a broadcast AM or FM radio to receive an amateur radio transmission unintentionally?
A. The receiver is unable to reject strong signals outside the AM or FM band
B. The microphone gain of the transmitter is turned up too high
C. The audio amplifier of the transmitter is overloaded
D. The deviation of an FM transmitter is set too low

How can fundamental overload of a non-amateur radio or TV receiver by an amateur signal be reduced or eliminated?
A. Block the amateur signal with a filter at the antenna input of the affected receiver
B. Block the interfering signal with a filter on the amateur transmitter
C. Switch the transmitter from FM to SSB
D. Switch the transmitter to a narrow-band mode

What should be the first step to resolve non-fiber optic cable TV interference caused by your amateur radio transmission?
A. Add a low-pass filter to the TV antenna input

B. Add a high-pass filter to the TV antenna input
C. Add a preamplifier to the TV antenna input
D. Be sure all TV feed line coaxial connectors are installed properly

Which of the following actions should you take if a neighbor tells you that your station's transmissions are interfering with their radio or TV reception?
A. Make sure that your station is functioning properly and that it does not cause interference to your own radio or television when it is tuned to the same channel
B. Immediately turn off your transmitter and contact the nearest FCC office for assistance
C. Install a harmonic doubler on the output of your transmitter and tune it until the interference is eliminated
D. All these choices are correct

What should you do if something in a neighbor's home is causing harmful interference to your amateur station?
A. Work with your neighbor to identify the offending device
B. Politely inform your neighbor that FCC rules prohibit the use of devices that cause interference
C. Make sure your station meets the standards of good amateur practice
D. All these choices are correct

What is the primary purpose of a dummy load?
A. To prevent the radiation of signals when making tests
B. To prevent over-modulation of your transmitter
C. To improve the radiation from your antenna
D. To improve the signal to noise ratio of your receiver

What does a dummy load consist of?
A. A high-gain amplifier and a TR switch
B. A non-inductive resistor and a heat sink
C. A low voltage power supply and a DC relay
D. A 50 ohm reactance used to terminate a transmission line.

6 DIFFERENT MODES

A Mode is the type of transmission that you are using. For example voice (which is most often called **phone** by hams, which is short for **radiotelephone**) is one common mode. Actually, there are two common voice modes that you will probably use. Those two common modes are **FM** (frequency modulation) and **SSB** (single sideband). One question asks about a disadvantage of FM over SSB (although it is also an advantage). That is that you can only receive one signal at a time.

SSB is a form of **AM (amplitude modulation)**. It is the type of voice modulation that is most often used for long-distance (sometimes called "weak signal") contacts on the VHF and UHF bands. SSB can be either Upper Sideband (USB) or Lower Sideband (LSB), and most transmitters allow you to switch between the two. Upper Sideband (USB) is normally used on 10 meters, and on VHF and UHF. Single sideband signals have narrower bandwidth than FM signals (in other words, they take up a smaller slice of frequencies than FM signals). A single sideband signal has a bandwidth of approximately 3 kHz. An FM signal of the type normally used on repeaters has a bandwidth between 10 and 15 kHz.

SSB is permitted on most amateur bands. For the test, you need to know that it is permitted on at least a portion of all of the bands above 50 MHz.

SSB signals are more difficult to tune in, and might take some practice. If the voice pitch of the signal sounds too high or too low, you need to adjust the frequency slightly. If you don't want to change the transmitter frequency, most transceivers have a **clarifier** or **RIT (receiver incremental tuning)** control. This control allows you to change the receiver frequency slightly to tune in the station.

The most inexpensive VHF transceivers are FM only. For weak-signal communication (in other words, being able to communicate more easily over long distances), you would want a transceiver that includes SSB and CW capabilities. In other words, you would want a **multi-mode** VHF transceiver.

FM is the type of modulation that is most commonly used for VHF and UHF voice repeaters. As noted above, the bandwidth is between 5 and 15 kHz. The **deviation** of the signal is basically the loudness of the audio signal, and this affects the bandwidth. The amount of deviation is determined by the amplitude of the modulating signal. As the deviation increases (in other words, the audio gets louder), the signal occupies more bandwidth. If the microphone gain of an FM transmitter is set to high, this can cause interference to nearby frequencies, because the deviation will be too large. As we discussed before, the part of the FM receiver that detects the FM signal is called the discriminator. Therefore, if you see a piece of equipment on the test that has a discriminator, then it must be an FM receiver.

One question on the test asks what might be the problem if your FM signal is distorting. The correct answer is that you might be talking too loudly. There's sometimes a tendency to talk louder when you know your signal is weak. But on FM, the signal will get through better if you talk more quietly into the microphone.

In any transmitter, FM or SSB, if the microphone gain control is set too high, the output signal may become distorted. If you are ever told that you are "**over deviating**" on FM, then you should talk **farther away** from the microphone. In many cases, this makes it easier to receive a weak signal, since it takes up less bandwidth.

CW stands for Continuous Wave, another name for a Morse code transmission. CW can be sent with a straight key, an electronic keyer, or a computer keyboard. CW is the type of emission with the narrowest bandwidth–in other words, it takes up the smallest slice of frequencies of all modes. The bandwidth of a CW signal is about 150 Hz.

If you know that FM, SSB, and CW are modes, this makes one of the questions very easy. If there is a switch marked "SSB/CW-FM," then this switch is obviously used to select the right mode. If the switch is on an amplifier, then the mode is actually selected by the transmitter. On an amplifier, that switch simply sets the amplifier to properly amplify that mode.

Amateurs can also transmit **fast-scan TV** transmissions (generally, the same as analog commercial television signals) on certain frequencies. These signals take up a large bandwidth of about 6 MHz. **NTSC** video is the type of video normally used.

Many receivers will have multiple choices for the **bandwidth** of the received signal. This allows you to select a bandwidth matching the mode, which can reduce noise or interference. For SSB reception, the optimum setting would be 2400 Hz.

Some **digital** modes are covered on the test. The first is **packet radio**. To use packet radio, you need a **terminal node controller** hooked between your computer and transceiver. VHF packet radio normally uses FM signals. All of the following may be part of a packet radio transmission: check sum which permits error detection, a header which contains the call sign of the station to which the information is being sent and automatic repeat request in case of error

Other digital modes use a sound card interface. For these modes, the sound card of the computer provides audio, which goes into the transceiver's microphone input. The sound card also converts received audio into digital form. The most common sound card mode is **PSK31**.

PSK31 is a low data-rate transmission method. PSK stands for **"phase shift keying"**. Other digital modes include **MFSK,** packet, IEEE 802.11, FT8, and JT65. The last two modes are part of a software package called **WSJT-X.** "WS" stands for "weak signal," and these are especially good for activities such as moonbounce (Earth-Moon-Earth), meteor scatter, or weak-signal beacons. **FT8** is currently very popular. You need to know that this mode can operate with a very low signal.

To use most of these digital modes, your computer sound card is connected to the radio to transfer transmit audio, receive audio, and push-to-talk. For example, the radio's speaker output is connected to the computer's microphone or line input.

An **Amateur Radio Mesh Network is** amateur-radio based data network using commercial Wi-Fi equipment with modified firmware.

A computer can be used in many ways as part of an amateur radio station. In addition to generating and decoding digital signals, it can be used for logging contacts and for sending and receiving CW.

A **beacon** is a type of station that transmits for the purpose of observing propagation or other experiments.

The test also covers space communications. Amateurs routinely communicate with stations in space, such as the International Space Station. They also routinely communicate with other amateurs on Earth, by using amateur satellites. This includes communication with amateurs in other countries. Any amateur who is licensed to transmit on the uplink frequency may operate through a satellite. And any amateur with a technician class license may communicate with an amateur station on the International Space Station on the 2 meter and 70 cm bands. Different satellites can use many modes, including SSB, FM, CW, and data.

Amateur satellites often transmit a beacon, which is a transmission from a space station that contains information about the satellite, such as its health and status.

To know where a satellite is located at a given time, you can use a satellite tracking program. These often show maps of the satellite's position, the times when the satellite passes, and the apparent frequency of the satellite at your location. To use one of these programs, you need to enter the satellite's Keplerian elements.

Because satellites are moving so fast, there is a **Doppler shift** in frequency, depending on the relative motion between the satellite and the earth station. **Spin fading** is caused by the rotation of the satellite and its antennas.

You need to know the abbreviation **LEO**. This stands for **Low Earth Orbit**. Various modes can be used to send signals to and from a satellite. One common mode is FM Packet. When you transmit to a satellite on the uplink frequency, you should use the minimum amount of power needed to complete the contact. This helps keep the satellite available for other users. If you use too much power, this can block others from using the satellite. A good way to judge that you are using the right amount of power is that your signal on the downlink should be about the same strength as the satellite's beacon.

Satellites generally transmit on one band and receive on another band. For example, if a satellite is in "**mode U/V**", this means that the uplink (the frequency you transmit on) is in the 70 cm band, and the downlink (the frequency you listen on) is in the 2 meter band.

Satellites use many modes, including CW, SSB, FM, and digital. Therefore, for the question asking about modes used by satellites, the correct answer is "all of these choices are correct."

There is one trick question about satellites that asks who is allowed to receive telemetry from a satellite. No license is required to receive signals, so the answer is that anyone can receive it.

Where may SSB phone be used in amateur bands above 50 MHz?

A. Only in sub-bands allocated to General class or higher licensees
B. Only on repeaters
C. In at least some segment of all these bands
D. On any band if the power is limited to 25 watts

What is the function of the SSB/CW-FM switch on a VHF power amplifier?
A. Change the mode of the transmitted signal
B. Set the amplifier for proper operation in the selected mode
C. Change the frequency range of the amplifier to operate in the proper segment of the band
D. Reduce the received signal noise

What is the effect of excessive microphone gain on SSB transmissions?
A. Frequency instability
B. Distorted transmitted audio
C. Increased SWR
D. All these choices are correct

Which of the following controls could be used if the voice pitch of a single-sideband signal seems too high or low?
A. The AGC or limiter
B. The bandwidth selection
C. The tone squelch
D. The receiver RIT or clarifier

Which of the following is a disadvantage of FM compared with single sideband?
A. Voice quality is poorer
B. Only one signal can be received at a time
C. FM signals are harder to tune
D. All these choices are correct

What is the advantage of having multiple receive bandwidth choices on a multimode transceiver?
A. Permits monitoring several modes at once by selecting a separate filter for each mode

B. Permits noise or interference reduction by selecting a bandwidth matching the mode
C. Increases the number of frequencies that can be stored in memory
D. Increases the amount of offset between receive and transmit frequencies

Which of the following receiver filter bandwidths provides the best signal-to-noise ratio for SSB reception?
A. 500 Hz
B. 1000 Hz
C. 2400 Hz
D. 5000 Hz

What can you do if you are told your FM handheld or mobile transceiver is over-deviating?
A. Talk louder into the microphone
B. Let the transceiver cool off
C. Change to a higher power level
D. Talk farther away from the microphone

What would cause your FM transmission audio to be distorted on voice peaks?
A. Your repeater offset is inverted
B. You need to talk louder
C. You are talking too loudly
D. Your transmit power is too high

Which of the following is a form of amplitude modulation?
A. Spread-spectrum
B. Packet radio
C. Single sideband
D. Phase shift keying (PSK)

What type of modulation is most commonly used for VHF packet radio transmissions?
A. FM or PM
B. SSB
C. AM
D. PSK

Which type of voice mode is most often used for long-distance (weak signal) contacts on the VHF and UHF bands?
A. FM
B. DRM
C. SSB
D. PM

Which type of modulation is most commonly used for VHF and UHF voice repeaters?
A. AM
B. SSB
C. PSK
D. FM or PM

Which of the following types of emission has the narrowest bandwidth?
A. FM voice
B. SSB voice
C. CW
D. Slow-scan TV

Which sideband is normally used for 10 meter HF, VHF and UHF single-sideband communications?
A. Upper sideband
B. Lower sideband
C. Suppressed sideband
D. Inverted sideband

What is a characteristic of single sideband (SSB) compared to FM?
A. SSB signals are easier to tune in correctly
B. SSB signals are less susceptible to interference
C. SSB signals have narrower bandwidth
D. All these choices are correct

What is the approximate bandwidth of a single sideband voice signal?
A. 1 kHz
B. 3 kHz
C. 6 kHz
D. 15 kHz

What is the approximate bandwidth of a VHF repeater FM phone signal?
A. Less than 500 Hz
B. About 150 kHz
C. Between 10 and 15 kHz
D. Between 50 and 125 kHz

What is the approximate bandwidth of AM fast-scan TV transmissions?
A. More than 10 MHz
B. About 6 MHz
C. About 3 MHz
D. About 1 MHz

What is the approximate maximum bandwidth required to transmit a CW signal?
A. 2.4 kHz
B. 150 Hz
C. 1000 Hz
D. 15 kHz

What is the FCC Part 97 definition of a beacon?
A. A government transmitter marking the amateur radio band edges
B. A bulletin sent by the FCC to announce a national emergency
C. A continuous transmission of weather information authorized in the amateur bands by the National Weather Service
D. An amateur station transmitting communications for the purposes of observing propagation or related experimental activities

Who may be the control operator of a station communicating through an amateur satellite or space station?
A. Only an Amateur Extra Class operator
B. A General class or higher licensee with a satellite operator certification
C. Only an Amateur Extra Class operator who is also an AMSAT member
D. Any amateur allowed to transmit on the satellite uplink frequency

What is the impact of using excessive effective radiated power on a satellite uplink?
A. Possibility of commanding the satellite to an improper mode
B. Blocking access by other users
C. Overloading the satellite batteries
D. Possibility of rebooting the satellite control computer

Which of the following is a way to determine whether your satellite uplink power is neither too low nor too high?
A. Check your signal strength report in the telemetry data
B. Listen for distortion on your downlink signal
C. Your signal strength on the downlink should be about the same as the beacon
D. All these choices are correct

What mode of transmission is commonly used by amateur radio satellites?
A. SSB
B. FM
C. CW/data

D. All of these choices are correct

Who may receive telemetry from a space station?
A. Anyone
B. A licensed radio amateur with a transmitter equipped for interrogating the satellite
C. A licensed radio amateur who has been certified by the protocol developer
D. A licensed radio amateur who has registered for an access code from AMSAT

Which of the following are provided by satellite tracking programs?
A. Maps showing the real-time position of the satellite track over the earth
B. The time, azimuth, and elevation of the start, maximum altitude, and end of a pass
C. The apparent frequency of the satellite transmission, including effects of Doppler shift
D. All of these answers are correct

Which of the following are inputs to a satellite tracking program?
A. The satellite transmitted power
B. The Keplerian elements
C. The last observed time of zero Doppler shift
D. All these choices are correct

Which amateurs may contact the International Space Station (ISS) on VHF bands?
A. Any amateur holding a General class or higher license
B. Any amateur holding a Technician class or higher license
C. Any amateur holding a General class or higher license who has applied for and received approval from NASA
D. Any amateur holding a Technician class or higher license who has applied for and received approval from NASA

What is a satellite beacon?
A. The primary transmit antenna on the satellite
B. An indicator light that shows where to point your antenna
C. A reflective surface on the satellite
D. A transmission from a satellite that contains status information

What telemetry information is typically transmitted by satellite beacons?
A. The signal strength of received signals
B. Time of day accurate to plus or minus 1/10 second
C. Health and status of the satellite
D. All of these choices are correct

What is Doppler shift in reference to satellite communications?
A. A change in the satellite orbit
B. A mode where the satellite receives signals on one band and transmits on another
C. An observed change in signal frequency caused by relative motion between the satellite and Earth station
D. A special digital communications mode for some satellites

What is meant by the statement that a satellite is operating in U/V mode?
A. The satellite uplink is in the 15 meter band and the downlink is in the 10 meter band
B. The satellite uplink is in the 70 centimeter band and the downlink is in the 2 meter band
C. The satellite operates using ultraviolet frequencies
D. The satellite frequencies are usually variable

What causes spin fading of satellite signals?
A. Circular polarized noise interference radiated from the sun
B. Rotation of the satellite and its antennas
C. Doppler shift of the received signal
D. Interfering signals within the satellite uplink band

What is a LEO satellite?
A. A sun synchronous satellite
B. A highly elliptical orbit satellite
C. A satellite in low energy operation mode
D. A satellite in low earth orbit

Which of the following is a digital communications mode?
A. Packet radio
B. IEEE 802.11
C. FT8
D. All these choices are correct

Which of the following operating activities is supported by digital mode software in the WSJT-X software suite?
A. Earth-Moon-Earth
B. Weak signal propagation beacons
C. Meteor scatter
D. All these choices are correct

How are the transceiver audio input and output connected in a station configured to operate using FT8?
A. To a computer running a terminal program and connected to a terminal node controller unit
B. To the audio input and output of a computer running WSJT-X software
C. To an FT8 conversion unit, a keyboard, and a computer monitor
D. To a computer connected to the FT8converter.com website

What signals are used in a computer-radio interface for digital mode operation?
A. Receive and transmit mode, status, and location
B. Antenna and RF power
C. Receive audio, transmit audio, and transmitter keying
D. NMEA GPS location and DC power

Which of the following connections is made between a computer and a transceiver to use computer software when operating digital modes?

A. Computer "line out" to transceiver push-to-talk
B. Computer "line in" to transceiver push-to-talk
C. Computer "line in" to transceiver speaker connector
D. Computer "line out" to transceiver speaker connector

What is FT8?
A. A wideband FM voice mode
B. A digital mode capable of low signal-to-noise operation
C. An eight channel multiplex mode for FM repeaters
D. A digital slow-scan TV mode with forward error correction and automatic color compensation

What type of transmission is indicated by the term "NTSC?"
A. A Normal Transmission mode in Static Circuit
B. A special mode for satellite uplink
C. An analog fast-scan color TV signal
D. A frame compression scheme for TV signals

What does the abbreviation PSK mean?
A. Pulse Shift Keying
B. Phase Shift Keying
C. Packet Short Keying
D. Phased Slide Keying

Which of the following is included in packet radio transmissions?
A. A check sum that permits error detection
B. A header that contains the call sign of the station to which the information is being sent
C. Automatic repeat request in case of error
D. All these choices are correct

What is CW?
A. A type of electromagnetic propagation
B. A digital mode used primarily on 2 meter FM
C. A technique for coil winding
D. Another name for a Morse code transmission

What is an electronic keyer?
A. A device for switching antennas from transmit to receive
B. A device for voice activated switching from receive to transmit
C. A device that assists in manual sending of Morse code
D. An interlock to prevent unauthorized use of a radio

7 DIFFERENT OPERATING ACTIVITIES

The test includes some questions about different kinds of operating activities.

One popular activity is **radio direction finding**, or **hidden transmitter hunts**. A transmitter hunt is a recreational activity, but radio direction finding may be used to locate the source of interference or jamming. A directional antenna is important for these activities.

Another popular activity is **contesting**, which involves contacting as many stations as possible during a specified period of time. If you contact another station in a contest, it is a good practice to send only the minimum information needed for proper identification and the contest exchange, often just a signal report and your location. In many VHF contests, the other station will want to know your **grid locator**, often known as your "grid square". This is two letters and two numbers which identify your geographic location. My grid locator, for example, is EN34.

Amateur radio can be used for radio control, which is the transmitting of telecommand signals to radio controlled models.

Repeaters can be linked together into a network. This just means that all of the repeaters are hooked together so that they repeat the signals received from the other repeaters. One way that repeaters are linked is through Internet Radio Linking Project (**IRLP**). IRLP is a technique to connect amateur radio systems, such as repeaters, via the Internet using Voice Over Internet Protocol (**VoIP**). VoIP is a method of delivering voice communications over the Internet using digital techniques.. To connect to a specific node, you use your radio's keypad to transmit the IRLP node ID (The tones generated by the keypad are called DTMF signals, which consist of a pair of two tones.) A station that links other stations to the internet is called a **gateway**.

One popular system of linked repeaters (and linked non-repeater stations) is **EchoLink**. You can access this system through repeaters, or through the Internet. In order to use EchoLink, you must register your call sign and provide proof of license. Because you are using a computer, one question points out that you don't need a radio.

Another way that repeaters can be linked together is through **DMR** (Digital Mobile Radio). In this system, users are in different "talkgroups," which is a way for groups of users to share a channel at different times without being heard by other users. To join a talk group, you would program your DMR radio with the group's ID or code. For one question, you need to know that the color code must match the color code of the repeater you are trying to access. "Code Plug" just means the access information for repeaters and talkgroups. For one question, remember that DMR allows two digital signals on the same channel. That question includes more details, but if you remember "two signals on one channel," you will get it right.

There is one question about a digital voice mode called D-STAR. You need to know that for D-STAR, you need to program the radio with your call sign before transmitting.

Some amateurs use **APRS** (Automatic Position Reporting System) to transmit their location and other information. A Global Positioning System (**GPS**) receiver is normally hooked up to the radio to automate this process. APRS can be used to provide real time tactical digital communications in conjunction with a map showing the locations of stations. APRS can be used to transmit GPS position, weather data, or text message.

When may an amateur station transmit without on-the-air identification?
A. When the transmissions are of a brief nature to make station adjustments
B. When the transmissions are unmodulated
C. When the transmitted power level is below 1 watt
D. When transmitting signals to control model craft

Which of the following methods is used to locate sources of noise interference or jamming?
A. Echolocation
B. Doppler radar
C. Radio direction finding
D. Phase locking

Which of these items would be useful for a hidden transmitter hunt?
A. Calibrated SWR meter
B. A directional antenna
C. A calibrated noise bridge
D. All of these choices are correct

What operating activity involves contacting as many stations as possible during a specified period?
A. Simulated emergency exercises
B. Net operations
C. Public service events
D. Contesting

Which of the following is good procedure when contacting another station in a contest?
A. Sign only the last two letters of your call if there are many other stations calling
B. Contact the station twice to be sure that you are in his log
C. Send only the minimum information needed for proper identification and the contest exchange
D. All of these choices are correct

What is a grid locator?
A. A letter-number designator assigned to a geographic location
B. A letter-number designator assigned to an azimuth and elevation
C. An instrument for neutralizing a final amplifier
D. An instrument for radio direction finding

Which of the following describes a linked repeater network?
A. A network of repeaters where signals received by one repeater are repeated by all the repeaters
B. A repeater with more than one receiver
C. Multiple repeaters with the same owner
D. A system of repeaters linked by APRS

Which of the following must be programmed into a D-STAR digital transceiver before transmitting?
A. Your call sign
B. Your output power
C. The codec type being used
D. All these choices are correct

What type of signaling uses pairs of audio tones?
A. DTMF
B. CTCSS
C. GPRS

D. D-STAR

What is required before using the EchoLink system?
A. Complete the required EchoLink training
B. Purchase a license to use the EchoLink software
C. Register your call sign and provide proof of license
D. All these choices are correct

Which of the following protocols enables an amateur station to transmit through a repeater without using a radio to initiate the transmission?
A. IRLP
B. D-STAR
C. DMR
D. EchoLink

How is over the air access to IRLP nodes accomplished?
A. By obtaining a password that is sent via voice to the node
B. By using DTMF signals
C. By entering the proper internet password
D. By using CTCSS tone codes

What name is given to an amateur radio station that is used to connect other amateur stations to the Internet?
A. A gateway
B. A repeater
C. A digipeater
D. A beacon

What is Voice Over Internet Protocol (VoIP)?
A. A set of rules specifying how to identify your station when linked over the internet to another station
B. A technique employed to "spot" DX stations via the internet
C. A technique for measuring the modulation quality of a transmitter using remote sites monitored via the internet

D. A method of delivering voice communications over the internet using digital techniques

What is the Internet Radio Linking Project (IRLP)?
A. A technique to connect amateur radio systems, such as repeaters, via the Internet using Voice Over Internet Protocol
B. A system for providing access to websites via amateur radio
C. A system for informing amateurs in real time of the frequency of active DX stations
D. A technique for measuring signal strength of an amateur transmitter via the Internet

How is a specific group of stations selected on a digital voice transceiver?
A. By retrieving the frequencies from transceiver memory
B. By enabling the group's CTCSS tone
C. By entering the group's identification code
D. By activating automatic identification

Which of the following best describes an amateur radio mesh network?
A. An amateur-radio based data network using commercial Wi-Fi equipment with modified firmware
B. A wide-bandwidth digital voice mode employing DMR protocols
C. A satellite communications network using modified commercial satellite TV hardware
D. An internet linking protocol used to network repeaters

What is a "talkgroup" on a DMR repeater?
A. A group of operators sharing common interests
B. A way for groups of users to share a channel at different times without hearing other users on the channel
C. A protocol that increases the signal-to-noise ratio when multiple repeaters are linked together
D. A net that meets at a specified time

How can you join a digital repeater's "talkgroup"?
A. Register your radio with the local FCC office

B. Join the repeater owner's club
C. Program your radio with the group's ID or code
D. Sign your call after the courtesy tone

What is the purpose of the color code used on DMR repeater systems?
A. Must match the repeater color code for access
B. Defines the frequency pair to use
C. Identifies the codec used
D. Defines the minimum signal level required for access

Which of the following describes DMR?
A. A technique for time-multiplexing two digital voice signals on a single 12.5 kHz repeater channel
B. An automatic position tracking mode for FM mobiles communicating through repeaters
C. An automatic computer logging technique for hands-off logging when communicating while operating a vehicle
D. A digital technique for transmitting on two repeater inputs simultaneously for automatic error correction

What does a DMR "code plug" contain?
A. Your call sign in CW for automatic identification
B. Access information for repeaters and talkgroups
C. The codec for digitizing audio
D. The DMR software version

Which of the following is an application of APRS?
A. Providing real-time tactical digital communications in conjunction with a map showing the locations of stations
B. Showing automatically the number of packets transmitted via PACTOR during a specific time interval
C. Providing voice over internet connection between repeaters
D. Providing information on the number of stations signed into a repeater

What kind of data can be transmitted by APRS?
A. GPS position data

B. Text messages
C. Weather data
D. All these choices are correct

8 OPERATING PROCEDURES

When you receive your license, your station will be issued a call sign. In the United States, a call sign will start with the letters A, K, N, or W. It will start with one or two letters (called the prefix), followed by a number, followed by up to three more letters (called the suffix). There is one question on the test about what a possible technician class call sign might look like, and the correct example is KF1XXX.

The test might ask how many licenses one person can have. The answer is one. (This is written in legalese, so you need to know that one person can have one "operator/primary station license grant.") The actual proof that you are licensed is the listing in the FCC database. The FCC no longer routinely mails out paper copies of licenses. Therefore, there is nothing to "display" or "have in your possession," so you know that those are the wrong answers.

There is one question about "vanity call signs," a call sign that you can select, if available. Any licensed amateur can participate in the vanity call sign program.

You are required to identify by sending your call sign every ten minutes during a contact, and then again at the end of the contact. If you are using phone (voice), this must be done in the English language (or, you can also satisfy the requirement by sending it in CW--Morse Code).

You must identify any time you transmit, even if you are just a test transmission. The normal rules apply during any kind of on-air test: You need to identify every ten minutes and at the end.

You can add things to your call sign, at the beginning or end. If you do so, this is called a self-assigned call sign indicator. It can't be in conflict with any other indicator specified by the FCC rules, and it can't be in conflict with the callsign prefix assigned to another country. The callsign would be written as KL7CC/W3, and on voice, you would say "KL7CC stroke W3" OR "KL7CC slant W3" OR "KL7CC slash W3". All of these are correct.

To call another station, you generally say the other station's callsign, and then identify with your call sign.

If you want to talk to any station, then you call CQ. "CQ" means "calling any station". (On many repeaters, instead of actually calling CQ, you just say your call sign followed by the word "monitoring.") If you hear another station calling CQ and want to answer, then you say their call sign, followed by your call sign.

Before picking a frequency to call CQ, you should listen first to make sure that no one is using the frequency, ask if the frequency is in use, and double check to make sure that it's a frequency you are allowed to use.

There are two "Q signals" covered on the test: QRM means that you are receiving interference from other stations. QSY means that you are changing frequency.

Many amateurs engage in various public service activities, including relaying messages (called "handling traffic"). It should be noted that the normal FCC rules apply, even if you are acting at the request of public service officials.

RACES and ARES are two organizations that provide communications during emergencies. RACES stands for "Radio Amateur Civil Emergency

Service", a service for communications during national emergencies governed by the FCC rules. It is a radio service using amateur stations for emergency management or civil defense communications.

When a group of amateurs are communicating, often for some public service purpose, this is referred to as a "net." One station, the Net Control Station (NCS), calls the net to order and directs communications between stations checking in. Unless you are reporting an emergency, you should transmit only when directed by the NCS.

The most important job of an amateur operator when handling emergency traffic messages is to pass along the message exactly as written, spoken, or as received. A formal message usually contains a **preamble**, which is the information needed to track the message. Part of the preamble is the **check**, which is simply a count of the number of words or word equivalents.

Third party communications means that a message is being sent to another station on behalf of a third party. (You—the control operator of your station--are the first party, the other station is the second party, and you are sending a message from someone else.)

You can transmit non-emergency third party messages to foreign countries, only if that station's country permits such communications. Most countries do not. It is important to remember that when a non-licensed person is speaking into the microphone of your station, this is a form of third-party message. Therefore, you cannot let that person speak to a station in another country, unless the U.S. has a third-party agreement with that country.

The only time you are ever allowed to operate on frequencies outside of your license class is when necessary in situations involving the immediate safety of life or property.

How many operator/primary station license grants may be held by any one person?
A. One
B. No more than two
C. One for each band on which the person plans to operate
D. One for each permanent station location from which the person plans to operate

What proves that the FCC has issued an operator/primary license grant?
A. A printed copy of the certificate of successful completion of examination
B. An email notification from the NCVEC granting the license
C. The license appears in the FCC ULS database
D. All these choices are correct

Which of the following is a valid Technician class call sign format?
A. KF1XXX
B. KA1X
C. W1XX
D. All these choices are correct

Who may select a desired call sign under the vanity call sign rules?
A. Only licensed amateurs with general or extra class licenses
B. Only licensed amateurs with an extra class license
C. Only an amateur licensee who has been licensed continuously for more than 10 years
D. Any licensed amateur

When are you required to transmit your assigned call sign?
A. At the beginning of each contact, and every 10 minutes thereafter
B. At least once during each transmission
C. At least every 15 minutes during and at the end of a communication
D. At least every 10 minutes during and at the end of a communication

Which of the following is an acceptable language to use for station identification when operating in a phone sub-band?

A. Any language recognized by the United Nations
B. Any language recognized by the ITU
C. The English language
D. English, French, or Spanish

What method of call sign identification is required for a station transmitting phone signals?
A. Send the call sign followed by the indicator RPT
B. Send the call sign using CW or phone emission
C. Send the call sign followed by the indicator R
D. Send the call sign using only phone emission

Which of the following formats of a self-assigned indicator is acceptable when identifying using a phone transmission?
A. KL7CC stroke W3
B. KL7CC slant W3
C. KL7CC slash W3
D. All of these choices are correct

Which of the following restrictions apply when a non-licensed person is allowed to speak to a foreign station using a station under the control of a licensed amateur operator?
A. The person must be a U.S. citizen
B. The foreign station must be in a country with which the U.S. has a third party agreement
C. The licensed control operator must do the station identification
D. All these choices are correct

What is an appropriate way to call another station on a repeater if you know the other station's call sign?
A. Say "break, break," then say the station's call sign
B. Say the station's call sign, then identify with your call sign

C. Say "CQ" three times, then the other station's call sign
D. Wait for the station to call CQ, then answer

How should you respond to a station calling CQ?
A. Transmit "CQ" followed by the other station's call sign
B. Transmit your call sign followed by the other station's call sign
C. Transmit the other station's call sign followed by your call sign
D. Transmit a signal report followed by your call sign

Which of the following is required when making on-the-air test transmissions?
A. Identify the transmitting station
B. Conduct tests only between 10 p.m. and 6 a.m. local time
C. Notify the FCC of the transmissions
D. All of these choices are correct

What is the meaning of the procedural signal "CQ"?
A. Call on the quarter hour
B. A new antenna is being tested (no station should answer)
C. Only the called station should transmit
D. Calling any station

Which of the following indicates that a station is listening on a repeater and looking for a contact?
A. "CQ CQ" followed by the repeater's call sign
B. The station's call sign followed by the word "monitoring"
C. The repeater call sign followed by the station's call sign
D. "QSY" followed by your call sign

Which Q signal indicates that you are receiving interference from other stations?
A. QRM
B. QRN

C. QTH
D. QSB

Which Q signal indicates that you are changing frequency?
A. QRU
B. QSY
C. QSL
D. QRZ

What should you do before calling CQ?
A. Listen first to be sure that no one else is using the frequency
B. Ask if the frequency is in use
C. Make sure you are authorized to use that frequency
D. All these choices are correct

What is the Radio Amateur Civil Emergency Service (RACES)?
A. A radio service using amateur frequencies for emergency management or civil defense communications
B. A radio service using amateur stations for emergency management or civil defense communications
C. An emergency service using amateur operators certified by a civil defense organization as being enrolled in that organization
D. All these choices are correct

What is RACES?
A. An emergency organization combining amateur radio and citizens band operators and frequencies
B. An international radio experimentation society
C. A radio contest held in a short period, sometimes called a "sprint"
D. An FCC part 97 amateur radio service for civil defense communications during national emergencies

Which of the following are typical duties of a Net Control Station?
A. Choose the regular net meeting time and frequency
B. Ensure that all stations checking into the net are properly licensed for operation on the net frequency
C. Call the net to order and direct communications between stations checking in

D. All these choices are correct

Which of the following is standard practice when you participate in a net?
A. When first responding to the net control station, transmit your call sign, name, and address as in the FCC database
B. Record the time of each of your transmissions
C. Unless you are reporting an emergency, transmit only when directed by the net control station
D. All these choices are correct

What is the definition of third party communications?
A. A message from a control operator to another amateur station control operator on behalf of another person
B. Amateur radio communications where three stations are in communications with one another
C. Operation when the transmitting equipment is licensed to a person other than the control operator
D. Temporary authorization for an unlicensed person to transmit on the amateur bands for technical experiments

What does the term "traffic" refer to in net operation?
A. Messages exchanged by net stations
B. The number of stations checking in and out of a net
C. Operation by mobile or portable stations
D. Requests to activate the net by a served agency

Which of the following is a characteristic of good traffic handling?
A. Passing messages exactly as received
B. Making decisions as to whether messages are worthy of relay or delivery
C. Ensuring that any newsworthy messages are relayed to the news media
D. All these choices are correct

Are amateur station control operators ever permitted to operate outside the frequency privileges of their license class?
A. No
B. Yes, but only when part of a FEMA emergency plan
C. Yes, but only when part of a RACES emergency plan
D. Yes, but only in situations involving the immediate safety of human life or protection of property

What information is contained in the preamble of a formal traffic message?
A. The email address of the originating station
B. The address of the intended recipient
C. The telephone number of the addressee
D. The information needed to track the message

What is meant by "check" in a radiogram header?
A. The number of words or word equivalents in the text portion of the message
B. The call sign of the originating station
C. A list of stations that have relayed the message
D. A box on the message form that indicates that the message was received and/or relayed

What is the Amateur Radio Emergency Service (ARES)?
A. A group of licensed amateurs who have voluntarily registered their qualifications and equipment for communications duty in the public service
B. A group of licensed amateurs who are members of the military and who voluntarily agreed to provide message handling services in the case of an emergency
C. A training program that provides licensing courses for those interested in obtaining an amateur license to use during emergencies
D. A training program that certifies amateur operators for membership in the Radio Amateur Civil Emergency Service

9 FCC RULES

The rules governing amateur radio are contained in **Part 97** of the rules of the FCC (Federal Communications Commission). The FCC regulates and enforces the amateur radio service in the United States. The Amateur Radio Service is intended for persons who are interested in radio technique solely with a personal aim and without pecuniary interest. There is one question that asks when the FCC rules don't apply. This is a trick question. The correct answer is that the FCC rules always apply.

The FCC rules begin by listing the purposes of Amateur Radio. Two of those purposes are included on the test: Advancing skills in the technical and communication phases of the radio art, and enhancing international goodwill.

You are allowed to communicate with amateurs in other countries. Your communications should be incidental to the purposes of the amateur service and remarks of a personal character. But you are not allowed to communicate with a country, if that country has notified the ITU that it objects to such communications. (However, there are currently no such countries.)

When you get your license, it will normally be valid for ten years. There is a two year grace period during which an expired license can be renewed. (However, you cannot transmit during this two year grace period.) Always keep your mailing address and your e-mail address current with the FCC. If mail is returned because it is undeliverable, or the FCC is unable to reach you by email, then the FCC can revoke your station license or suspend your operator license. When you pass your test, you can begin operating a transmitter as soon as your license appears in the FCC ULS database, which you can access online. You do not need to wait for the license to actually come in the mail. After you have your license, if you pass the test for General or Extra, you can begin to operate with your new license privileges if you have a CSCE (certificate of successful completion of element). When you do this, you are required to transmit /AE or /AG after your call sign. (Someone who has an old Novice license who upgrades to Technician can do the same by using /KT. Even though this is very rare, it is mentioned in one of the correct answers.)

The FCC currently issues three kinds of new amateur licenses: Technician, General, and Extra. If you forget the names, remember that there are only three. All of the wrong answers list four types.

Part 97 of the FCC rules includes a number of definitions of terms you need to know for the test:

Automatic control is the situation when the control operator is not at the control point. (If that definition doesn't make much sense yet, just memorize it, since it will make more sense when we talk about who a "control operator" is. Two examples of automatic control that are on the test are an APRS network digipeater or a repeater operation.

An **auxiliary station** transmits signals over the air from a remote receive site to a repeater for retransmission.

An **amateur station** is a station in an Amateur Radio Service consisting of the apparatus necessary for carrying on radio communications.

Broadcasting is any transmission intended for reception by the general public. Amateurs are not allowed to broadcast. And many new hams need to be careful not to use the word "broadcast" to describe what they do. Amateurs are allowed to transmit, but they are not allowed to broadcast. In general, Amateurs are allowed to make two-way transmissions, not one-way. The only exceptions are for code practice, information bulletins, telecommand or telemetry, or transmissions necessary to provide emergency communications.

Harmful interference is interference which seriously degrades, obstructs, or repeatedly interrupts a radio communication service operating in accordance with the Radio Regulations. Willful interference is never permitted. If two stations are interfering because they are on the same frequency, they should use common sense and negotiate a way that they can both use it.

Remote control is the situation when the control operator is not at the station location, but can indirectly manipulate the operating adjustments of a station. The control operator must be at the control point at all times. Do not confuse this with radio control of a model craft. When the phrase "remote control" is used on the test, this means operating a radio from a remote location, such as controlling the radio through the Internet.

A **Repeater station** is the type of amateur station that simultaneously retransmits the signal of another amateur station on a different channel or channels. If a repeater station inadvertently retransmits communications that violate the FCC rules, then the control operator of the originating station is responsible.

A **space station** is an amateur station located more than 50 km above the Earth's surface.

Telemetry is a one-way transmission of measurements at a distance from the measuring instrument .

You can never use codes or ciphers on the air to obscure the meaning of your transmission. There is only one exception to this rule, and the exception is on the test. You can use codes or ciphers only when transmitting control commands to space stations or radio control craft.

You can also never transmit music, but there is also one rare exception, and the exception is on the test. It is OK to transmit music only if it is incidental to an authorized retransmission of manned spacecraft communications.

Obscene or indecent words or language are never allowed.

Normally, Amateur Radio cannot be used for any money-making purpose. There is an exception which allows you to notify other amateurs of equipment that you have for sale or trade. The equipment must be normally used in an amateur station, and you can't do this on a regular basis.

In general, the control operator of an amateur station may never receive compensation for operating the station. One of the rare exceptions, which is on the test, is if the communication is incidental to classroom instruction at an educational institution.

There is only one situation where an amateur station may transmit signals related to broadcasting, program production, or news gathering. That is when the communications relate to the immediate safety of human life or protection of property, and no other means is available. Amateurs are not allowed to retransmit any entertainment programming from a commercial or public TV or radio station.

You are allowed to make brief transmissions to make station adjustments.

The control operator of a station is, as the name implies, the person operating the controls. To be a control operator, you must be an FCC licensed amateur (in other words, you appear in the ULS database). Also, an alien who is authorized for reciprocal operation because he is licensed in his home country may be the control operator. A station must have a control operator whenever the station is transmitting. The control operator must be designated by the station licensee. (In other words, if you are operating your own station, you designate yourself as the control operator.)

The transmitting privileges of the station are determined by the class of operator license of the control operator. For example, after you get your technician license, even if you operate someone else's station, you can only transmit on technician frequencies.

If the control operator and the station licensee are two different people, then both of them are responsible for the proper operation of the station. The FCC presumes that the control operator and the station licensee are the same person, in the absence of any documentation to the contrary in station records.

The control point is the location where the control operator function is performed. There is one confusing question on the test that asks what type of control is being used when transmitting using a handheld radio. The fact that the radio is handheld is irrelevant. Since the person operating the radio is obviously right there at the radio, this is an example of local control, and local control is the correct answer.

We talked previously about station identification requirements. As you remember, whenever you are transmitting, you need to give your call sign every ten minutes, and at the end.

Sometimes, in addition to your call sign, if you are participating in some sort of public service activity, you might use a "tactical call sign" to identify your role in the activity. For example, one station might be identified as "Race Headquarters". This is a "tactical call sign". Again, if you are using a "tactical call sign", you still need to use your normal FCC call sign every ten minutes.

In order to obtain a club station license, a radio club must have at least four members. The trustee of the club station license is the only person who can request a vanity call sign for the club.

As a station licensee, you need to make your station records available for FCC inspection any time they are requested by an FCC representative.

The only rule about power covered by the test (other than the one about radio control models, which was one watt) is the rule that says you must use the minimum transmitter power necessary to carry out the desired communication.

Which of the following is a purpose of the Amateur Radio Service as stated in the FCC rules and regulations?
A. Providing personal radio communications for as many citizens as possible
B. Providing communications for international non-profit organizations
C. Advancing skills in the technical and communication phases of the radio art
D. All of these choices are correct

Which agency regulates and enforces the rules for the Amateur Radio Service in the United States?
A. FEMA
B. Homeland Security
C. The FCC
D. All of these choices are correct

What is the FCC Part 97 definition of a space station?
A. Any satellite orbiting the earth
B. A manned satellite orbiting the earth
C. An amateur station located more than 50 km above the Earth's surface
D. An amateur station using amateur radio satellites for relay of signals

When is willful interference to other amateur radio stations permitted?
A. To stop another amateur station which is breaking the FCC rules

B. At no time
C. When making short test transmissions
D. At any time, stations in the Amateur Radio Service are not protected from willful interference

For which licenses classes are new licenses currently available from the FCC?
A. Novice, Technician, General, Advanced
B. Technician, Technician Plus, General, Advanced
C. Novice, Technician Plus, General, Advanced
D. Technician, General, Amateur Extra

When do FCC rules NOT apply to the operation of an amateur station?
A. When operating a RACES station
B. When operating under special FEMA rules
C. When operating under special ARES rules
D. FCC rules always apply

What types of international communications are an FCC-licensed amateur radio station permitted to make?
A. Communications incidental to the purposes of the Amateur Radio Service and remarks of a personal character
B. Communications incidental to conducting business or remarks of a personal nature
C. Only communications incidental to contest exchanges; all other communications are prohibited
D. Any communications that would be permitted by an international broadcast station

From which of the following locations may an FCC-licensed amateur station transmit?
A. From within any country that belongs to the International Telecommunication Union
B. From within any country that is a member of the United Nations
C. From anywhere within International Telecommunication Union (ITU) Regions 2 and 3

D. From any vessel or craft located in international waters and documented or registered in the United States

What may happen if the FCC is unable to reach you by email?
A. Fine and suspension of operator license
B. Revocation of the station license or suspension of the operator license
C. Revocation of access to the license record in the FCC system
D. Nothing; there is no such requirement

Which of the following can result in revocation of the station license or suspension of the operator license?
A. Failure to inform the FCC of any changes in the amateur station following performance of an RF safety environmental evaluation
B. Failure to provide and maintain a correct email address with the FCC
C. Failure to obtain FCC type acceptance prior to using a home-built transmitter
D. Failure to have a copy of your license available at your station

What is the normal term for an FCC-issued amateur radio license?
A. Five years
B. Life
C. Ten years
D. Eight years

What is the grace period for renewal if an amateur license expires?
A. Two years
B. Three years
C. Five years
D. Ten years

How soon after passing the examination for your first amateur radio license may you transmit on the amateur radio bands?
A. Immediately on receiving your Certificate of Successful Completion of Examination (CSCE)
B. As soon as your operator/station license grant appears on the ARRL website

C. As soon as your operator/station license grant appears in the FCC's license database
D. As soon as you receive your license in the mail from the FCC

If your license has expired and is still within the allowable grace period, may you continue to transmit on the amateur radio bands?
A. Yes, for up to two years
B. Yes, as soon as you apply for renewal
C. Yes, for up to one year
D. No, you must wait until the license has been renewed

With which countries are FCC-licensed amateur radio stations prohibited from exchanging communications?
A. Any country whose administration has notified the International Telecommunication Union (ITU) that it objects to such communications
B. Any country whose administration has notified the American Radio Relay League (ARRL) that it objects to such communications
C. Any country banned from such communications by the International Amateur Radio Union (IARU)
D. Any country banned from making such communications by the American Radio Relay League (ARRL)

When is it permissible to transmit messages encoded to obscure their meaning?
A. Only during contests
B. Only when transmitting certain approved digital codes
C. Only when transmitting control commands to space stations or radio control craft
D. Never

Under which of the following circumstances are one-way transmissions by an amateur station prohibited?
A. In all circumstances

B. Broadcasting

C. International Morse Code Practice

D. Telecommand or transmissions of telemetry

Under what conditions is an amateur station authorized to transmit music using a phone emission?
A. When incidental to an authorized retransmission of manned spacecraft communications
B. When the music produces no spurious emissions
C. When transmissions are limited to less than three minutes per hour
D. When the music is transmitted above 1280 MHz

When may amateur radio operators use their stations to notify other amateurs of the availability of equipment for sale or trade?
A. Never
B. When the equipment is not the personal property of either the station licensee, or the control operator, or their close relatives
C. When no profit is made on the sale
D. When selling amateur radio equipment and not on a regular basis

What, if any, are the restrictions concerning transmission of language that may be considered indecent or obscene?
A. The FCC maintains a list of words that are not permitted to be used on amateur frequencies
B. Any such language is prohibited
C. The ITU maintains a list of words that are not permitted to be used on amateur frequencies
D. There is no such prohibition

In which of the following circumstances may the control operator of an amateur station receive compensation for operating that station?
A. When the communication is related to the sale of amateur equipment by the control operator's employer
B. When the communication is incidental to classroom instruction at an educational institution

C. When the communication is made to obtain emergency information for a local broadcast station
D. All of these choices are correct

When may amateur stations transmit information in support of broadcasting, program production, or news gathering, assuming no other means is available?
A. When such communications are directly related to the immediate safety of human life or protection of property
B. When broadcasting communications to or from the space shuttle
C. Where noncommercial programming is gathered and supplied exclusively to the National Public Radio network
D. Never

How does the FCC define broadcasting for the Amateur Radio Service?
A. Two-way transmissions by amateur stations
B. Any transmission made by the licensed station
C. Transmission of messages directed only to amateur operators
D. Transmissions intended for reception by the general public

When may an amateur station transmit without a control operator?
A. When using automatic control, such as in the case of a repeater
B. When the station licensee is away and another licensed amateur is using the station
C. When the transmitting station is an auxiliary station
D. Never

Who must designate the station control operator?
A. The station licensee
B. The FCC
C. The frequency coordinator
D. The ITU

What determines the transmitting frequency privileges of an amateur station?
A. The frequency authorized by the frequency coordinator
B. The frequencies printed on the license grant
C. The highest class of operator license held by anyone on the premises
D. The class of operator license held by the control operator

What is an amateur station control point?
A. The location of the station's transmitting antenna
B. The location of the station transmitting apparatus
C. The location at which the control operator function is performed
D. The mailing address of the station licensee

Which of the following is an example of automatic control?
A. Repeater operation
B. Controlling the station over the Internet
C. Using a computer or other device to automatically send CW
D. Using a computer or other device to automatically identify

When the control operator is not the station licensee, who is responsible for the proper operation of the station?
A. All licensed amateurs who are present at the operation
B. Only the station licensee
C. Only the control operator
D. The control operator and the station licensee

Which of the following are required for remote control operation?
A. The control operator must be at the control point
B. A control operator is required at all times
C. The control operator must indirectly manipulate the controls
D. All these choices are correct

Which of the following is an example of remote control as defined in Part 97?
A. Repeater operation
B. Operating the station over the Internet
C. Controlling a model aircraft, boat or car by amateur radio
D. All of these choices are correct

Who does the FCC presume to be the control operator of an amateur station, unless documentation to the contrary is in the station records?
A. The station custodian
B. The third party participant
C. The person operating the station equipment
D. The station licensee

How often must you identify with your FCC-assigned call sign when using tactical call signs such as "Race Headquarters"?
A. Never, the tactical call is sufficient
B. Once during every hour
C. At the end of each communication and every ten minutes during a communication
D. At the end of every transmission

When, under normal circumstances, may a Technician class licensee be the control operator of a station operating in an Amateur Extra Class band segment?
A. At no time
B. When designated as the control operator by an Amateur Extra Class licensee
C. As part of a multi-operator contest team
D. When using a club station whose trustee holds an Amateur Extra Class license

What type of amateur station simultaneously retransmits the signal of another amateur station on a different channel or channels?
A. Beacon station
B. Earth station
C. Repeater station
D. Message forwarding station

Who is accountable should a repeater inadvertently retransmit communications that violate the FCC rules?

A. The control operator of the originating station
B. The control operator of the repeater
C. The owner of the repeater
D. Both the originating station and the repeater owner

Which of the following is a requirement for the issuance of a club station license grant?
A. The trustee must have an Amateur Extra class operator license grant
B. The club must have at least four members
C. The club must be registered with the American Radio Relay League
D. All of these choices are correct

When must the station licensee make the station and its records available for FCC inspection?
A. At any time ten days after notification by the FCC of such an inspection
B. At any time upon request by an FCC representative
C. Only after failing to comply with an FCC notice of violation
D. Only when presented with a valid warrant by an FCC official or government agent

Which of the following applies when two stations transmitting on the same frequency interfere with each other?
A. The stations should negotiate continued use of the frequency
B. Both stations should choose another frequency to avoid conflict
C. Interference is inevitable, so no action is required
D. Use subaudible tones so both stations can share the frequency

10 SAFETY

There are a number of important safety issues that are covered on the test. The first of these is electric shock. Obviously, current flowing through the body is a hazard, because it can heat tissue, it disrupts the electrical functions of cells, and it can cause involuntary muscle contractions.

Here are some ways to guard against electrical shock:

Use three-wire cords and plugs for all AC powered equipment.

Connect all AC powered station equipment to a common safety ground. The green wire on a standard three-wire electrical plug needs to be hooked to ground. The black wire is the "hot" connection.

Use mechanical interlocks on high-voltage circuits.

Even if a power supply is turned off and disconnected, you could still receive an electric shock from stored charge in large capacitors. And when measuring high voltages with a voltmeter, it's important to make sure that the insulation on the probes and wires can handle that much voltage.

If you ever build equipment that is powered by 120 volt electricity, you must include a fuse or circuit breaker in series with the AC "hot" conductor.

Lightning protection is also important. If your antenna is struck by lightning, this could cause a fire. In the case of a coaxial cable running into your station, you should install a lighting arrester on a grounded panel near where feed lines enter the building. The correct way to ground a tower is to use a separate eight-foot ground rod for each tower leg, bonded to the tower and each other.

Ground wires on a tower for lightning protection must be short and direct. Sharp bends must be avoided. If you have more than one ground rod, they should be bonded together with heavy wire or conductive strap. The best option is flat copper strap. Grounding requirements for towers and antennas are governed by local electrical codes.

Working on antennas often involves working off the ground, and can be dangerous for that reason. In addition, you must always be aware of power lines. Contact with a power line could easily prove fatal.

When climbing a tower, always use a climbing harness. Never climb a tower without a helper or observer.

You must never climb a crank-up tower unless it is in the fully retracted position.

When putting up a tower, look for and stay clear of overhead electrical wires. And whenever setting up any antenna, make sure that it is clear of power lines. The rule is that you shouldn't put it up, unless no part of it will come closer than 10 feet to power wires if it falls down unexpectedly. Don't attach antennas to utility poles, because the antenna could come into contact with high-voltage wires.

There is one question about guy wires. You need to know that a safety wire through a turnbuckle is to prevent it from loosening from vibration.

Finally, the test has some questions about exposure to RF radiation. The station licensee is responsible for ensuring that no person is exposed to RF energy above the FCC exposure limits. RF radiation is called non-ionizing radiation, and it is different from ionizing radiation (radioactivity). One main difference is that RF radiation does not have sufficient energy to cause genetic damage. However, it is possible that RF exposure might have some health effects, and there are rules governing RF exposure. The amount of exposure depends on a number of factors. These include the frequency and power, the distance from the antenna, and the radiation pattern of the antenna. The human body absorbs some frequencies more than others, and the distance, power, and radiation pattern affect the amount of exposure.

Another factor is the duty cycle of the transmitter. This means the percentage of time that it is transmitting. This is an important factor because it affects the average amount of radiation that people receive. For example, if the transmitter is on 100% of the time, this is twice as much average exposure than it would be if the transmitter was only on 50% of the time, and you can use twice as much power with the same exposure level.

In some circumstances, you are required to do an RF exposure evaluation of your station. There are various ways to do the evaluation. These include all of the following:

By calculation based on FCC OET Bulletin 65

By calculation based on computer modeling

By measurement of field strength using calibrated equipment

If your station is over the limits, one way to fix the problem is to relocate the antenna. You need to re-evaluate the station whenever an item of equipment is changed.

One of the questions asks which frequency has the lowest value for Maximum Permissible Exposure limit. The correct answer is 50 MHz.

If someone accidentally touches your antenna while you are transmitting, they could get a very painful RF burn.

Which of the following precautions should be taken when measuring high voltages with a voltmeter?
A. Ensure that the voltmeter has very low impedance
B. Ensure that the voltmeter and leads are rated for use at the voltages to be measured
C. Ensure that the circuit is grounded through the voltmeter
D. Ensure that the voltmeter is set to the correct frequency

What health hazard is presented by electrical current flowing through the body?
A. It may cause injury by heating tissue
B. It may disrupt the electrical functions of cells
C. It may cause involuntary muscle contractions
D. All these choices are correct

In the United States, what circuit does black wire insulation indicate in a three-wire 120 V cable?
A. Neutral
B. Hot
C. Equipment ground
D. Black insulation is never used

What is a good way to guard against electrical shock at your station?
A. Use three-wire cords and plugs for all AC powered equipment
B. Connect all AC powered station equipment to a common safety ground
C. Install mechanical interlocks in high-voltage circuits
D. All these choices are correct

Where should a lightning arrester be installed in a coaxial feed line?
A. At the output connector of a transceiver
B. At the antenna feed point
C. At the ac power service panel
D. On a grounded panel near where feed lines enter the building

Which of the following is good practice when installing ground wires on a tower for lightning protection?
A. Put a drip loop in the ground connection to prevent water damage to the ground system
B. Make sure all ground wire bends are right angles
C. Ensure that connections are short and direct
D. All these choices are correct

What hazard exists in a power supply immediately after turning it off?
A. Circulating currents in the dc filter
B. Leakage flux in the power transformer
C. Voltage transients from kickback diodes
D. Charge stored in filter capacitors

What is required when climbing an antenna tower?
A. Have sufficient training on safe tower climbing techniques
B. Use appropriate tie-off to the tower at all times
C. Always wear an approved climbing harness
D. All these choices are correct

Under what circumstances is it safe to climb a tower without a helper or observer?
A. When no electrical work is being performed
B. When no mechanical work is being performed
C. When the work being done is not more than 20 feet above the ground
D. Never

Which of the following is an important safety precaution to observe when putting up an antenna tower?
A. Wear a ground strap connected to your wrist at all times
B. Insulate the base of the tower to avoid lightning strikes
C. Look for and stay clear of any overhead electrical wires
D. All of these choices are correct

What is the minimum safe distance from a power line to allow when installing an antenna?
A. Add the height of the antenna to the height of the power line and multiply by a factor of 1.5
B. The height of the power line above ground
C. 1/2 wavelength at the operating frequency
D. Enough so that if the antenna falls, no part of it can come closer than 10 feet to the power wires

Which of the following is an important safety rule to remember when using a crank-up tower?
A. This type of tower must never be painted
B. This type of tower must never be grounded
C. This type of tower must not be climbed unless retracted or mechanical safety locking devices have been installed
D. All of these choices are correct

Which is a proper grounding method for a tower?
A. A single four-foot ground rod, driven into the ground no more than 12 inches from the base
B. A ferrite-core RF choke connected between the tower and ground
C. A connection between the tower base and a cold water pipe
D. Separate eight-foot ground rods for each tower leg, bonded to the tower and each other

Why should you avoid attaching an antenna to a utility pole?
A. The antenna will not work properly because of induced voltages
B. The 60 Hz radiations from the feed line may increase the SWR
C. The antenna could contact high-voltage power lines
D. All these choices are correct

Which of the following is true when installing grounding conductors used for lightning protection?

A. Use only non-insulated wire
B. Wires must be carefully routed with precise right-angle bends
C. Sharp bends must be avoided
D. Common grounds must be avoided

Which of the following establishes grounding requirements for an amateur radio tower or antenna?
A. FCC Part 97 Rules
B. Local electrical codes
C. FAA tower lighting regulations
D. UL recommended practices

What is the purpose of a safety wire through a turnbuckle used to tension guy lines?
A. Secure the guy line if the turnbuckle breaks
B. Prevent loosening of the turnbuckle from vibration
C. Provide a ground path for lightning strikes
D. Provide an ability to measure for proper tensioning

What type of radiation are radio signals?
A. Gamma radiation
B. Ionizing radiation
C. Alpha radiation
D. Non-ionizing radiation

How does the allowable power density for RF safety change if duty cycle changes from 100 percent to 50 percent?
A. It increases by a factor of 3
B. It decreases by 50 percent
C. It increases by a factor of 2
D. There is no adjustment allowed for lower duty cycle

Who is responsible for ensuring that no person is exposed to RF energy above the FCC exposure limits?
A. The FCC
B. The station licensee
C. Anyone who is near an antenna

D. The local zoning board

At which of the following frequencies does maximum permissible exposure have the lowest value?
A. 3.5 MHz
B. 50 MHz
C. 440 MHz
D. 1296 MHz

What factors affect the RF exposure of people near an amateur station antenna?
A. Frequency and power level of the RF field
B. Distance from the antenna to a person
C. Radiation pattern of the antenna
D. All of these choices are correct

Why do exposure limits vary with frequency?
A. Lower frequency RF fields have more energy than higher frequency fields
B. Lower frequency RF fields do not penetrate the human body
C. Higher frequency RF fields are transient in nature
D. The human body absorbs more RF energy at some frequencies than at others

Which of the following is an acceptable method to determine that your station complies with FCC RF exposure regulations?
A. By calculation based on FCC OET Bulletin 65
B. By calculation based on computer modeling
C. By measurement of field strength using calibrated equipment
D. All of these choices are correct

What hazard is created by touching an antenna during a transmission?
A. Electrocution
B. RF burn to skin
C. Radiation poisoning
D. All these choices are correct

Which of the following actions can reduce exposure to RF radiation?
A. Relocate antennas
B. Relocate the transmitter
C. Increase the duty cycle
D. All these choices are correct

How can you make sure your station stays in compliance with RF safety regulations?
A. By informing the FCC of any changes made in your station
B. By re-evaluating the station whenever an item of equipment is changed
C. By making sure your antennas have low SWR
D. All of these choices are correct

Why is duty cycle one of the factors used to determine safe RF radiation exposure levels?
A. It affects the average exposure of people to radiation
B. It affects the peak exposure of people to radiation
C. It takes into account the antenna feed line loss
D. It takes into account the thermal effects of the final amplifier

What is the definition of duty cycle during the averaging time for RF exposure?
A. The difference between the lowest power output and the highest power output of a transmitter
B. The difference between the PEP and average power output of a transmitter
C. The percentage of time that a transmitter is transmitting
D. The percentage of time that a transmitter is not transmitting

How does RF radiation differ from ionizing radiation (radioactivity)?
A. RF radiation does not have sufficient energy to cause chemical changes in cells and damage DNA
B. RF radiation can only be detected with an RF dosimeter

C. RF radiation is limited in range to a few feet
D. RF radiation is perfectly safe

What should be done to all external ground rods or earth connections?
A. Waterproof them with silicone caulk or electrical tape
B. Keep them as far apart as possible
C. Bond them together with heavy wire or conductive strap
D. Tune them for resonance on the lowest frequency of operation

Which of the following conductors is preferred for bonding at RF?
A. Copper braid removed from coaxial cable
B. Steel wire
C. Twisted-pair cable
D. Flat copper strap

11 MISCELLANEOUS

There are a few questions on the test which don't fit in well with the material in the other chapters. To make sure that you ace the test, those items are included here.

A **frequency coordinator** is a local volunteer (either an individual or an organization) that recommends transmit and receive channels and other parameters for auxiliary and repeater stations. The frequency coordinator is selected by amateur operators in a local or regional area whose stations are eligible to be auxiliary or repeater stations.

A **band plan** is a voluntary guideline for using different modes or activities within an amateur band.

Transmitting and receiving on the same frequency is called **simplex** communication. If you are close enough to another station to communicate directly by simplex, then you should consider doing so, rather than tying up a repeater.

When you identify your station using phone, the FCC rules encourage you to use a **phonetic alphabet**. It's also a good idea to use a standard phonetic alphabet when sending messages with proper names and unusual words.

To reduce RF current flowing on the shield of an audio cable, you could use a **ferrite choke**.

Reports of garbled, distorted, or unintelligible transmissions might be the result of RF feedback in a transmitter or receiver. Distortion of an FM signal might also be caused by being off frequency.

An **antenna analyzer** is an instrument used to determine if an antenna is resonant at a desired frequency.

ARQ is an error correction method in which the receiving station detects errors and sends a request for retransmission.

One question asks which kinds of amateur stations can automatically retransmit other stations. The correct answer is auxiliary, repeater, or space stations.

Shielded wire is often used to prevent coupling of unwanted signals.

To determine how long a battery will run a piece of equipment, look at the number of amp hours on the battery, and divide that by the number of amps the equipment uses. This will give you the number of hours.

Which of the following entities recommends transmit/receive channels and other parameters for auxiliary and repeater stations?
A. Frequency Spectrum Manager appointed by the FCC
B. Volunteer Frequency Coordinator recognized by local amateurs
C. FCC Regional Field Office
D. International Telecommunications Union

Who selects a Frequency Coordinator?
A. The FCC Office of Spectrum Management and Coordination Policy
B. The local chapter of the Office of National Council of Independent Frequency Coordinators
C. Amateur operators in a local or regional area whose stations are eligible to be auxiliary or repeater stations
D. FCC Regional Field Office

What is a band plan, beyond the privileges established by the FCC?
A. A voluntary guideline for using different modes or activities within an amateur band
B. A mandated list of operating schedules
C. A list of scheduled net frequencies
D. A plan devised by a club to indicate frequency band usage

What term describes an amateur station that is transmitting and receiving on the same frequency?
A. Full duplex
B. Diplex
C. Simplex
D. Multiplex

Why are simplex channels designated in the VHF/UHF band plans?
A. So stations within range of each other can communicate without tying up a repeater
B. For contest operation
C. For working DX only
D. So stations with simple transmitters can access the repeater without automated offset

What technique is used to ensure that voice messages containing unusual words are received correctly?
A. Send the words by voice and Morse code
B. Speak very loudly into the microphone
C. Spell the words using a standard phonetic alphabet

D. All these choices are correct

What do the FCC rules state regarding the use of a phonetic alphabet for station identification in the Amateur Radio Service?
A. It is required when transmitting emergency messages
B. It is encouraged
C. It is required when in contact with foreign stations
D. All these choices are correct

Which of the following could you use to cure distorted audio caused by RF current flowing on the shield of a microphone cable?
A. Band-pass filter
B. Low-pass filter
C. Preamplifier
D. Ferrite choke

What might be a problem if you receive a report that your audio signal through an FM repeater is distorted or unintelligible?
A. Your transmitter is slightly off frequency
B. Your batteries are running low
C. You are in a bad location
D. All these choices are correct

What is the result of tuning an FM receiver above or below a signal's frequency?
A. Change in audio pitch
B. Sideband inversion
C. Generation of a heterodyne tone
D. Distortion of the signal's audio

What is a symptom of RF feedback in a transmitter or transceiver?
A. Excessive SWR at the antenna connection
B. The transmitter will not stay on the desired frequency
C. Reports of garbled, distorted, or unintelligible transmissions
D. Frequent blowing of power supply fuses

Which of the following instruments can be used to determine if an antenna is resonant at the desired operating frequency?
A. A VTVM
B. An antenna analyzer
C. A Q meter
D. A frequency counter

What is an ARQ transmission system?
A. A special transmission format limited to video signals
B. A system used to encrypt command signals to an amateur radio satellite
C. An error correction method in which the receiving station detects errors and sends a request for retransmission
D. A method of compressing data using autonomous reiterative Q codes prior to final encoding

What types of amateur stations can automatically retransmit the signals of other amateur stations?
A. Auxiliary, beacon, or Earth stations
B. Earth, repeater, or space stations
C. Beacon, repeater, or space stations
D. Repeater, auxiliary, or space stations

Which of the following is a common reason to use shielded wire?
A. To decrease the resistance of DC power connections
B. To increase the current carrying capability of the wire
C. To prevent coupling of unwanted signals to or from the wire
D. To couple the wire to other signals

How can you determine the length of time that equipment can be powered from a battery?
A. Divide the watt-hour rating of the battery by the peak power consumption of the equipment
B. Divide the battery ampere-hour rating by the average current draw of the equipment
C. Multiply the watts per hour consumed by the equipment by the battery power rating

D. Multiply the square of the current rating of the battery by the input resistance of the equipment

ABOUT THE AUTHOR

Richard Clem, WØIS, has been a licensed amateur since 1974, having previously held the calls WNØMEB and WBØMEB. His previous works include a multi-band antenna construction article published in QST. He is married to the illustrator, Yippy Clem, KCØOIA, and resides in St. Paul, Minnesota. He is an attorney in private practice.

Made in the USA
Middletown, DE
31 October 2023

41706887R00086